Climate Change and Environmental Ethics

Climate Change and Environmental Ethics

Ved P. Nanda
editor

ROUTLEDGE
Routledge
Taylor & Francis Group
New York London

First published 2011 by Transaction Publishers

Published 2017 by Routledge
711 Third Avenue, New York, NY 10017, USA
2 Park Square, Milton Park, Abingdon, Oxon OX14 4RN

First issued in hardback 2017

Routledge is an imprint of the Taylor & Francis Group, an informa

Library of Congress Catalog Number: 2010043862

Library of Congress Cataloging-in-Publication Data

Nanda, Ved P.
Climate change and environmental ethics / Ved P. Nanda.
p. cm.
Includes bibliographical references and index.
ISBN 978-1-4128-1459-1 (alk. paper)
1. Climatic changes. 2. Environmental ethics. I. Title.

QC903.N355 2011
179'.1--dc22

2010043862

ISBN 13: 978-1-4128-4967-8 (pbk)
ISBN 13: 978-1-1385-2066-0 (hbk)

Contents

Introduction

Ved P. Nanda

> *The changing climate is pushing many Earth systems towards critical thresholds that will alter regional and global environmental balances and threaten stability at multiple scales. Alarmingly, we may have already passed tipping points that are irreversible within the time span of our current civilization.*
>
> *—UNEP Year Book 2009, 21*

Decision makers are giving serious attention to this UNEP warning and its assertion that climate change "has long since ceased to be a scientific curiosity" (ibid.). On July 8, 2009, at their summit meeting in L'Aquila, Italy, leaders of the Group of Eight[1] leading nations agreed to cut emissions of heat-trapping gases by 80 percent by 2050 (G-8, Global Issues). They also accepted a ceiling of two degrees Celsius for total global warming (ibid.). However, there was no agreement on shorter-term targets to cut emissions or on a firm amount of aid to set aside for developing countries. Also, their declaration leaves it to individual nations to decide the emissions baseline from which they will count their reductions (ibid.).

On the following day at the G-8 summit, US President Barack Obama addressed the Major Economies Forum on Energy and Climate (MEF), attended by seventeen developed and developing countries,[2] stating:

> Every nation on this planet is at risk. And just as no one nation is responsible for climate change, no one nation can address it alone. That is why, back in April, I convened this forum of the world's major economies—responsible for more than three-quarters of the world's carbon pollution. And it is why we have gathered again here today (Major Economies Forum 2009).

Leaders of the MEF did not set any firm targets or timetable because there had been no consensus among the developed countries on either their targets for shorter-term emissions reductions or on the resources and technology that they would provide to the developing countries for mitigation and adaptation, hence the unwillingness of the developing nations to accept firm targets. However, the MEF leaders did agree "to work between now and Copenhagen [December 2009] to identify a global goal for substantially reducing global emissions by 2050, recognizing the science indicating that global warming should not exceed two degrees Celsius above pre-industrial levels" (ibid.). There also was general agreement on advancing "clean energy technology," on funding to assist developing countries, and on advancing adaptation efforts (ibid.).

The response to the L'Aquila climate-change declaration was mixed. US officials considered it as progress (Jonathan Weisman 2009). The Indian Prime Minister's special envoy on climate change, Shyam Saran, concurred, stating that "the declaration was positive and forward looking and would send the right kind of message, though the specifics would have to be negotiated under the United Nations Framework Convention on Climate Change leading to the Copenhagen summit" (N. Ravi 2009). But a spokesman for Oxfam International, who summed up the view of several environmental groups present at L'Aquila, said: "The G-8 might have agreed to avoid cooking the planet by more than two degrees, but they made no attempt to turn down the heat anytime soon" (Jonathan Weisman 2009). Subsequently, the Copenhagen Summit, held from December 7 to December 19, 2009, failed to reach accord on specifics, thus postponing the decision for a future meeting (Copenhagen Accord 2009).

A major change on the international landscape concerning international responses to the challenge of climate change is the new US policy under the Obama administration. In June 2009, the United States government's Global Change Research Program released a report which confirmed, based on observations, that "warming of the climate is unequivocal" (US June 2009 Report). It added: "The global warming observed over the past 50 years is due primarily to human-induced emissions of heat-trapping gases. These emissions come mainly from the burning of fossil fuels (coal, oil, and gas), with additional contributions from the clearing of forests and agricultural activities)" (ibid.).

The new policy reverses the Bush administration's rejection of the Kyoto Protocol, which had been based primarily on Kyoto's requiring of firm commitments by nations to reduce carbon emissions. In contrast, President Obama committed the US to take a leading role in tackling climate change as he addressed an audience of tens of thousands outside the medieval Prague Castle in Prague, Czech Republic, on April 5, 2009: "Together, we must confront climate change by ending the world's dependence on fossil fuels, by tapping the power of new sources of energy like the wind and sun, and calling upon all nations to do their part. And I pledge to you that in this global effort, the United States is now ready to lead" (Obama in Prague 2009). When he initially launched the Forum on March 28, 2009, its objective was described as to

> facilitate a candid dialogue among key developed and developing countries, help generate the political leadership necessary to achieve a successful outcome at the UN climate change negotiations that will convene this December in Copenhagen, and advance the exploration of concrete initiatives and joint ventures that increase the supply of clean energy while cutting greenhouse gas emissions (White House March 28, 2009).

Although there is a broad consensus that climate change presents the international community with a formidable challenge, and notwithstanding that the US is now on board to tackle this thorny issue, progress on all fronts—prevention, mitigation, and adaptation—has been alarmingly slow. The next section briefly presents the nature and scope of the challenge and why the responses have not been robust. In the last analysis, nations have to take effective steps and there has to be concerted international cooperative action, which translates into decision makers and all constituents in a nation-state—local and municipal units, businesses, and non-governmental organizations—taking appropriate action so that the optimum outcome is reached. This obviously cannot happen without a changed attitude on the part of human beings, which necessarily calls for a new ethical commitment.

This book, *Climate Change and Environmental Ethics*, collects first-rate research and thinking of scholars from multiple disciplines—ethics, ecology, ecological philosophy, economics, political science, history, and international law. The group also included leaders of non-governmental organizations. They came together under the auspices of the Toda Institute in Honolulu in the fall of 2008, each bringing her/his special perspective. The focus of the deliberations was on exploring solutions and on the need to change the human mindset. Although the Copenhagen meeting was unable to reach an accord on future action by states to address the challenge of climate change, they must grapple with the issue to help ensure that we do not reach the tipping point, so that the planet not only survives but remains healthy for future generations. The contributions in this volume should benefit those making decisions for the future.

Enormity of the Challenge

New research published since the 2007 release of the Fourth Assessment Report of the Intergovernmental Panel on Climate Change (IPCC 2007) paints a bleaker picture of the planet's environment. The IPCC assessed that global warming is "unequivocal" and is evidenced by increased glacial and snow melting, increased air and ocean temperatures, and rising sea levels (IPCC 2007 Summary). According to the IPCC, this global warming coincides with anthropogenic contributions of greenhouse gases (GHG), such as carbon dioxide, methane, and nitrous oxide, and changes in land uses (ibid.).

However, a 2009 study by climate scientists at the Massachusetts Institute of Technology (MIT) indicates that global warming could be much worse than previously estimated. As reported:

> The most comprehensive modeling yet carried out on the likelihood of how much hotter the Earth's climate will get in this century shows that without rapid and massive action, the problem will be about twice as severe as previously estimated six years ago—and could be even worse than that (MIT 2009).

In May 2010, the US National Academy of Sciences' National Research Council issued its comprehensive study, America's Climate Choices, reporting that "[p]rojections of future climate change anticipate an additional warming of 2.0 to 11.5°F (1.1 to 6.4°C) over the 21st century, on top of the 1.4°F already observed over the past 100 years" (National Academy of Sciences 2010).

UNEP's 2009 Year Book reports that the ice melt in the Arctic, the Antarctic, and Greenland is much worse than the IPCC had projected, and that it is formally to be attributed to human influence. The UNEP Report notes that researchers used place-specific gridded data sets and simulations from four different climate models, concluding that "human activities have already caused significant warming in both polar regions with likely consequences for indigenous communities, biological systems, ice-sheet mass balance, and global sea levels" (UNEP Year Book 2009, 22). Citing a 2008 Nansen Environmental and Remote Sensing Center study, the Report states:

> For the second year in a row, there was an ice-free channel in the Northwest Passage through the islands of northern Canada. But this year also saw the opening of the Northern Sea Route along the Arctic Siberian coast. The two passages have probably not been open simultaneously since before the last ice age, some 100,000 years ago … Theoretically, in 2008 the Arctic ice cap could have been circumnavigated (ibid.).

An April 2009 study by NASA scientists confirmed this finding by reporting that "ice cover on the North Pole is shrinking and getting thinner, too. That's both a sign and a cause of continued rising temperatures …" (NASA April 2009). A similar result was also revealed in a 2008 study by the European Space Agency (ESA 2008).

As to the Greenland and Antarctica ice sheet loss, the report cites a 2008 study stating, "New findings in 2008 revealed that the flow into the ocean of the Jakobshavn and Isbrae glacier in western Greenland, one of the most important routes for ice loss, has doubled since 1997" (UNEP Year Book 2009, 23). The European Space Agency reported in April 2009 on the risk of the ice bridge supporting the Wilkins Ice Shelf, which connects it to the Charcot and Latady Islands, partly breaking away from the Antarctic Peninsula (ESA April 2009). Another NASA study reported in January 2009 that West Antarctica is now also warming, according to a climate researcher at the University of Washington in Seattle who led the study (NASA January 2009).

A disturbing scenario is the potential for the physical breakup of the ice sheets of Greenland and Antarctica, which could result in the rise of global sea levels to far exceed the last IPCC assessment forecast of up to 60 centimeters by 2100. What if the sea level were to rise by one meter by the end of the century? It would certainly create havoc globally for coastal states, as the Maldives, along with Kiribati and Tuvalu in the Pacific and the island of Sunderbans in the Bay of Bengal, would be submerged. And, as the UN High Commissioner for Refugees has observed:

> Sinking island states present one of the most dramatic scenarios of the impact of climate change. The entire populations of low-lying States such as the Maldives, Tuvalu, Kiribati and the Marshall Islands may in future be obliged to leave their own country as a result of climate change. Moreover, the existence of their State as such may be threatened. Entire populations of affected states could thus become stateless (UNHCR June 2009, 1).

Based upon research since the IPCC Assessment Forecast, UNEP states that a much larger rise is possible and indeed probable. It cites several studies predicting a rise between .8 and 1.5 meters as the likely outcome by the end of this century, which would displace around 100 million people in Asia. The Report concludes that

> research in 2008 indicates that sea level rise—from thermal expansion, mountain glacier retreat, and ice sheet melt—is likely to be much greater and to arrive much sooner than believed even two years ago. No matter how quickly climate change is

mitigated, sea level will rise. So, efforts to adapt to rising seas are more urgent than ever (ibid., 25).

Thus, the IPCC chairman, Dr. Rajendra K. Pachauri, has aptly called upon politicians to respond to new research showing that the sea level rise is likely to be far worse than the prior UN projections. "They should certainly respond to a worst-case scenario—even if there is only a small risk that it becomes reality—because it will have terrible consequences. It is quite common risk management" (Pachauri March 2009).

As to the Amazon rainforest, which contains a quarter of the world's species, the Report notes that it could be on a climatic edge in 2008, citing a prediction by the Hadley Center at Britain's Meteorological Office that "the Amazon will be close to a crucial tipping point. Beyond that point, the almost daily rainfall that sustains the jungle will become less dependable, soils will dry out, and much of the forest will die" (UNEP Year Book 2009, 26).

Among several adverse consequences of climate change, the following three will be noted here for illustrative purposes: the expansion in the size of the tropical zone, the potential disappearance of key coral reefs, and the drying up of big rivers. Agence France Presse reported an Australian study concluding that climate change is rapidly expanding the size of the tropical zone, which threatens to bring disease and drought to heavily populated areas (AFP 2009).

A study by the international conservation group WWF warned in May 2009 that 40 percent of reefs in the Coral Triangle, comprising a network of reefs among Indonesia, the Philippines, Malaysia, Papua New Guinea, the Solomon Islands, and East Timor, have already been lost. This area contains more than 76 percent of the world's reef-building coral species and 35 percent of its coral reef fish species, and supports the livelihoods of nearly 100 million people. The report was based on 300 published studies and called upon developed countries to cut their carbon emissions to 40 percent below 1990 levels by 2020 and developing economies to cut their emissions by at least 30 percent below current levels to prevent a catastrophe (WWF May 2009).

An April 2009 study by the National Center for Atmospheric Research, published in the American Meteorological Society's *Journal of Climate*, reveals that big rivers in some of the world's most populated regions are drying up. These include the Yellow River in China, the Ganges in India, the Niger in West Africa, and the Colorado in the Southwestern United States, and climate change is a likely cause. The result could be a potential threat to future supplies of food and water. According to the lead NCAR scientist on the study, "Reduced runoff is increasing the pressure on freshwater resources in much of the world, especially with more demand for water as population increases." He added, "Freshwater being a vital resource, the downward trends are a great concern" (NCAR April 2009). At a May 2010 regional conference of the Food and Agriculture Organization (FAO) for Africa in Luanda, Angola, a background document addressed the implications of climate change for African food security (Climate Change Implications 2010). After studying African vulnerability to adverse effects of climate changes, it concluded that the likely impact will "reduce crop yields and exacerbate the risk of food insecurity in Africa (Climate Change Implications 2010, 15).

Is all this potential adverse impact of climate change leading to disastrous consequences? The UNEP Report is not sanguine. The following two statements are pertinent: "With possibilities of collapsing ice sheets, methane bubbling out of permafrost, desiccated rainforest ecosystems, and sporadic ocean circulation patterns, concern is growing that Earth's life-support systems are approaching thresholds that contain tipping points" (UNEP Year Book 2009, 28); and citing two studies from the Proceedings of the National Academy of Sciences, the Report concludes that:

> [T]he evidence suggests that we may be within a few years of crossing tipping points with potential to disrupt seasonal weather patterns that support the agriculture activities of half the human population, diminish carbon sinks in the oceans and on land, and destabilize major ice sheets that could introduce unanticipated rates of sea level rise within the 21st century (ibid., 29).

If humanity faces dire consequences unless it resolutely acts to take effective mitigation actions and to undertake appropriate adaptation measures, why has there been such reluctance and resistance to do so? Obviously, there is not the political will to act firmly, notwithstanding the urgency. There is a sharp divide between the developed and developing countries on how to operationalize the "common but differentiated responsibilities" principle. Developing countries demand that major industrialized nations, which have primarily caused the problem, commit to steep cuts in carbon emissions in the short term and provide the necessary resources and technology so that countries without the wherewithal to undertake mitigation and adaptation measures are able to participate in the solution to this crisis. Many developed countries, on the other hand, seek firm commitments and timetables from the developing countries, without which, they argue, no international agreement can be reached. The impasse at L'Aquila underscored this divide, as did the later Copenhagen meeting.

Under the current state-centered international system, nations make agreements and commitments. The United Nations Framework Convention on Climate Change, the Kyoto Protocol, and any successor treaty that might emerge attest to this elementary proposition. International law will certainly provide the necessary framework with appropriate norms, institutions, and processes as the vehicle to move toward the needed action. Future negotiations will sort out the details. Decision makers will, one hopes, take into account the accumulated wisdom of other disciplines—both natural sciences and social sciences—as they devise the most appropriate mechanisms to address the challenge. However, the solution to the climate change crisis cries out for a change in the human mindset, without which the efforts cannot succeed. This, indeed, calls for a new ecological ethic.

Perspectives to Meet the Challenge

The scholars gathered in Honolulu acknowledged the gravity of the challenge and brought their disciplines as well as their personal perspectives to the deliberations. Their contributions here are divided into five parts. The first part comprises essays by four leading environmental philosophers. J. Baird Callicott, a professor of philosophy and religion studies at the University of North Texas and co-editor in chief of the *Encyclopedia of Environmental Ethics and Philosophy,* who is known as the leading contemporary exponent of Aldo Leopold's land ethic, studies Leopold's monumental contribution as anticipating the Gaia Hypothesis—that the Earth or biosphere is, as a whole, a living being—and thus "a worthy object of moral respect."

The next essay is by Professor Robin Attfield, an environmental philosopher and ethicist from Cardiff University. He asserts that biocentrism is "intrinsically superior as an ethic, and significantly strengthens the case for strong international action on climate change." In the third essay, ecological philosopher Freya Mathews of LaTrobe University, Australia, studies moral ambiguities in the politics of climate change, arguing for a bio-inclusive environmentalism in the era of climate change. And in the fourth essay, Professor Thomas Heyd of the University of Victoria, B.C., Canada argues that by taking "note of the cultural frameworks that enable and constrain our actions" we may be able to address climate change.

In Part Two, a political scientist, an ecological economist, and two professors of philosophy propose new governance and economics approaches to meet the challenge. In the opening essay, Sheila D. Collins, who directs the graduate program in Public Policy and International Affairs at William Patterson University in New Jersey, offers refreshing insights as she argues for a new paradigm of global governance and economy. An ecologist/economist at Southern Oregon University, John Gutrich proposes an ecological economics approach to move humanity towards sustainable development as we face global climate change, for he considers concerted international effort and a renewed environmental ethic as prerequisites to safeguard our global atmosphere. Professors Andrew Brennan and Y.S. Lo, also at LaTrobe University, suggest the need for bold initiatives to "liberate people from the addictions of affluenza," which they find as the main cause of inadequate responses to climate change as well as the current financial meltdown.

In Part Three, two international lawyers provide the perspective of international law and human rights and a biologist considers the need for a new and strong ethical and moral commitment on the part of the developed countries in order to ameliorate poverty as envisioned in the UN's Millennium Development Goals. Ved P. Nanda, Professor of International Law at the University of Denver and author of a leading treatise on international environmental law, suggests that both conventional and customary international law from human rights and environmental law perspectives "can play a robust role as appropriate mechanisms are crafted to support developing countries in their response to the adverse impacts of climate change."

Professor Laura Westra, both a philosopher and an international law scholar, provides an in-depth study of "linking intergenerational and intragenerational rights in ecojustice." She calls for the manifestation of the necessary political will to realize the urgency of securing both these rights. Professor Alicia Villamizar of the Simon Bolivár University in Caracas, Venezuela, who is a scholar of both biology and ecology, contends that "a dark combination of poverty and climate change poses a formidable challenge to reaching the MDG [Millennium Development Goals], and harshly exposes the weakness of the environmental morality of the planet." She calls on the developed world to provide assistance to avert the potential catastrophes of hurricanes, floods, and extreme droughts that the developing countries face.

Part Four comprises contributions by two civil society leaders. Katrina S. Rogers, Associate Dean for the Doctoral Program at Fielding

Graduate University, Santa Barbara, California, who also directs the university's Institute for Social Innovation, studies the case of The Nature Conservancy to make us understand the use of "soft" power by non-governmental organizations. She considers such power "as an explanatory framework in which to discuss the possible evolution of a more ethical system of collaboration and attraction, rather than coercion and force." Dune Lankard is a native Alaskan Eyak who has initiated several not-for-profit organizations and organized several "for-profit sustainable and sensible businesses." He also helped lead the community effort to convince Exxon to settle the Exxon Valdez Oil Spill litigation in an out-of-court settlement that established a $900 million "restoration fund" for mitigation of the spill's impacts and to help preservation and restoration efforts. He details a riveting story of this episode and his activism.

In the final part, Part Five, two case studies are presented—one by an environmental geologist writing on Cameroon and the other by an environmental scholar on Morocco. Professor Samuel Ayonghe, Vice Dean of the Faculty of Science at the University of Buea, Cameroon, who has coordinated several national and international research projects with the United Nations Environmental Program (UNEP), studies trends of climate change in Cameroon, in comparison with global trends, and describes the government's initiatives, some of which he considers as an expression of renewed environmental and political ethics. He recommends several specific policies that the government should adopt for mitigation and adaptation.

Professor and former Director of the Moroccan Ministry of the Environment Taja Balafrej evaluates the situation in Morocco, suggesting that his country become a leader in innovative clean technologies. For Morocco to succeed in its efforts to undertake projects combining mitigation and adaptation it must become self-reliant; but in the short term it will have to rely on international cooperation and the support of developed countries.

Conclusion

What distinguishes the collection of these contributions from much recent work on climate change are two of its special features—one is the multi-disciplinary backgrounds of the scholars, combined with their stellar experiences and the wisdom with which they have expressed not simply their philosophy and theory, but also their suggestions for concrete, specific action in practical terms. The second special niche to be filled by this volume is the overarching theme of the need for a renewed environmental ethic that permeates these disparate but interconnected pieces.

To illustrate, in Part One, Callicott sets the tone as he invokes Aldo Leopold's Gaian speculations to call for a "collective socio-cultural response in the form of policy, regulation, treaty, and law." Powerful cases for biocentrism are provided by Mathews and Attfield, each delving into philosophy as well as rigorous analysis and offering pragmatic solutions. And Heyd's reflection on how important it is to transform "the conceptions, values, practices, priorities, and material culture that make up the prevailing cultural framework" is invaluable as a point of departure toward the future in which we must "develop the robust coping behaviors and resilience that we need to effectively face the challenges posed by climate change." After all, scientific expertise alone will not solve environmental problems, for human values are equally important as complementary tools.

Accustomed as we are to the prevailing political and economic governance systems, Part Two compels us to do some out-of-the-box thinking as we are asked to consider the views of historians and scientists in order to rethink the concept of the nation-state system and the capitalist economy, a concept we have not revisited for several centuries. We are also urged to compare neo-classical economics with ecological economics and to shed our current obsession with consumerism. Instead, we are invited to consider new paradigms of global governance and economy.

While not a departure from the traditional study of the role of international law and policy in addressing critical issues facing humanity, Part Three provides a sharper focus on linking international law, human rights, and development. By presenting the challenge in stark terms and calling upon decision makers to use the tools that international law provides—norms, institutions, and processes—Nanda and Westra present a compelling case for international cooperation and the application of "common but differentiated responsibilities" to assist developing countries.

Civil society's significant role as an agent of change is now widely appreciated. The contributions in Part Five convincingly drive home how organizations and individuals can have a spectacular impact. The Nature Conservancy, with its annual budget of $900 million and a staff of 3,200 in thirty countries and all fifty US states, has resources exceeding those of many countries' conservation budgets and of most NGOs. Its impressive accomplishments will sound incredible to many readers, but, as Rogers explains, that is soft power in action, a phenomenon which has yet to be fully comprehended. And how a passionate, driven, and skilled individual can bring about monumental changes is Dune Lankard's fascinating story.

Finally, case studies of Morocco and Cameroon by Balafrej and Ayonghe vividly call to mind the hardships and obstacles many developing countries face as they lack effective institutions and adequate human and financial resources to tackle climate change. How a few committed individuals are working tenaciously to grapple with difficult issues and to find creative solutions is highlighted in these two final case studies.

In a nutshell, this collection presents a conversation about alternative ways of conceiving of our relation to the natural world, as one of my colleagues has aptly expressed it. A spirit of international cooperation and collaboration is sorely needed to meet the challenge and this book is the Toda Institute's modest contribution, which, we hope will compel the reader to think afresh about our values and more specifically about our ethical responsibilities regarding climate change.

Notes

1. The eight nations are the United States, Canada, France, Germany, Italy, Japan, Russia, and the United Kingdom.
2. Participants were Australia, Brazil, Canada, China, the European Union, France, Germany, India, Indonesia, Italy, Japan, Korea, Mexico, Russia, South Africa, the United Kingdom, and the United States. Denmark, in its capacity as President of the 2009 Conference of the Parties to the UN Framework Convention on Climate Change, and the United Nations also participated.

References

America's Climate Choices 2010. National Academy of Sciences, America's Climate Choices – Advancing the Science of Climate Change, Report in Brief, May 2010.

Climate Change Implications: Food and Agriculture Organization of the United Nations, Twenty-Sixth Regional Conference for Africa, Luanda, Angola, 030=-07 May 2010, Climate Change Implications for Food Security and Natural Resources Management for Africa, FAO Doc. ARC/10/8, May 2010.

AFP 2009. Climate change increasing width of tropics – With zone expanding 500Km in 25 years, droughts & epidemics threaten heavily populated areas, July 7, 2009, available at http://epaper.timesofindia.com/Default/Scripting/ArticleWin.asp?From=Archive&Source=Page&Skin=TOI&BaseHref=CAP/2009/07/07&PageLabel=17&EntityId=Ar01700&ViewMode=HTML&GZ=T.

Copenhagen Accord 2009. UNFCCC (Framework Convention on Climage Change), Report of the Conference of the Parites on its fifteenth session, held in Copenhagen from 7 to 19 December 2009, Addendum, Part Two, U.N. Doc. FCCC/CP/2009/11/Add.1, 30 March 2010, Decision 2/CP.15, 4.

ESA 2008. European Space Agency, Arctic ice on the verge of another all-time low, 28 August 2008, available at http://www.esa.int/esaCP/SEMCKX0SAKF_index_2.html.

ESA April 2009. European Space Agency, Collapse of the ice bridge supporting Wilkins Ice Shelf appears imminent, 3 April 2009, available at http://www.esa.int/esaCP/SEMD07EH1TF_index_2.html

IPCC 2007 Summary. IPCC, 2007: Summary for Policy Makers *in* Climate Change 2007: The Physical Science Basis. Contribution of Working Group I to the Fourth Assessment Report of the Intergovernmental Panel on Climate Change 5 (Cambridge University Press, Cambridge, UK, and New York, USA), available at http://ipcc/wg1.ucar.

Major Economies Forum 2009. White House, Office of the Press Secretary, *Meeting the International Clean Energy and Climate Change Challenges*, L'Aquila, Italy, July 9, 2009, *available at* http://www.whitehouse.gov/the_press_office/Fact-Sheet-Meeting-the-International-Clean-Energy-and-Climate-Change-Challenges/.

MIT 2009. David Chandler, MIT News Office, Climate change odds much worse than thought: New analysis shows warming could be double previous estimates, May 19, 2009, available at http://web.mit.edu/newsoffice/2009/roulette-0519.html.

NASA January 2009. NASA, Satellites Confirm Half-Century of West Antarctic Warming, January 21, 2009, available at www.nasa.gov.tpics/earth/features/warming_antarctica.html.

NASA April 2009. NASA, Ice, Ice Maybe: NASA Reports Greater Arctic Ice Melt, April 6, 2009

NCAR April 2009. Water Levels Dropping in Some major Rivers as Global Climate Changes, April 21, 2009, available at http://www.ucar.edu/news/releases/2009/flow.jsp.

Obama in Prague 2009. Remarks by President Barack Obama, Hradcany Square, Prague, Czech Republic, April 5, 2009, available at http://www.whitehouse.gov/the_press_office/Remarks-By-President-Barack-Obama-In-Prague-As-Delivered.

Pachauri March 2009. Rajendra K. Pachauri, IPCC-chairman to politicians: Respond to worst-case scenarios – Head of the UN's Intergovernmental Panel on Climate Change urges politicians to respond to new scientific climate research, March 3, 2009, available at http://en.cop15.dk/news/view+news?newsid=868.

N. Ravi 2009. N. Ravi, *A strong political message on climate change: India,* The Hindu Online, July 10, 2009.

UNEP Year Book 2009: UNEP Year Book 2009 (2009) – New Science and Developments in Our Changing Environment, New York, United Nations.

UNHCR June 2009. UN High Commissioner for Refugees, Submission: Climate Change and Statelessness: An Overview, presented to the 6th session of the Ad Hoc Working Group on Long-Term Cooperative Action (AWG-LCA 6) under the UN Framework Convention on Climate change (UNFCCC) 1 to 12 June 2009, Bonn, Germany, available at http://unfccc.ing/resource/docs/2009/smsn/igo/048.pdf.

US June 2009 Report. U.S. Global Change Research Program, Global Climate Change Impacts in the United States, Executive Summary 9, Washington DC, June 2009, available at http://www.globalchange.gov.

Jonathan Weisman 2009. Jonathan Weisman, *Summit leaders in climate deal: G-8 agrees to cut emissions of heat-trapping gases 80%, but wider group drops numbers*, Wall Street Journal (Europe), July 9, 2009, p. 1 (quoting Todd Stern, Chief US Climate-Change Negotiator).

White House March 28, 2009. Office of the Press Secretary, White House, President Obama Announces Launch of the Major Economies Forum on Energy and Climate, Press Release, March 28, 2009, available at http://www.whitehouse.gov/the_press_office/President-Obama-Announces-Launch-of-the-Major-Economies-Forum-on-Energy-and-Climate.

WWF May 2009. Reuters, Coral Triangle at risk from climate change – WWF, May 12, 2009, available at http://www.reuters.com/article/latestCrisis/idUSJAK465609.

Part I

Philosophy

1

Toward an Earth Ethic: Aldo Leopold's Anticipation of the Gaia Hypothesis

J. Baird Callicott

Aldo Leopold's 1949 "The Land Ethic" is seminal in academic environmental ethics and the environmental ethic of choice among professional conservationists and environmentalists. After sixty years, the sciences (evolutionary biology and ecology) that inform the land ethic have undergone much change. The land ethic can be revised to accommodate changes in its scientific foundations, but it cannot be scaled up to meet the challenge of global climate change. Fortunately, given the prominent place of Leopold in all circles environmental, he also faintly sketched an Earth Ethic in a paper written in 1923, which was published posthumously in 1979. The Earth Ethic is informed less by ecology and evolutionary biology than by biogeochemistry and anticipates the Gaia Hypothesis, viz., that the Earth (or biosphere) is, as a whole, a living being. If so, Leopold thought it a worthy object of moral respect.

Aldo Leopold, the Prophet

Aldo Leopold is often called a prophet, mainly because he was warning of the onslaught of an environmental crisis—although not by that name—more than a decade before Rachel Carson (1962) and Stewart Udall (1963) sounded the alarm.[1] Heralding impending doom is, after all, one of the signal things prophets do. Further, Leopold's seminal essay, "The Land Ethic," anticipated, by more than two decades, the emergence of formal, academic environmental ethics—which came on the scene in the early 1970s.[2] And that too is prophetic—albeit of a good thing, not a bad. One can find scattered remarks in the works of Henry David Thoreau and especially John Muir that intimate a need for a new, non-anthropocentric moral relationship on the part of humans with nature. Leopold's "The Land Ethic," however, is the first sustained expression of such a need and the first outline of how such a relationship might be understood—at least in the Western intellectual tradition—and put into practice.

The book in which "The Land Ethic" appears, *A Sand County Almanac*, is, moreover, often called the bible of the contemporary environmental movement (Stegner 1989). That is no accident. Leopold carefully crafted its style to give a subtle biblical ring and cadence to its words; and so—through his primary audience's Sunday-schooled ear—the book subliminally exudes an almost divine authority (Tallmadge 1989). Part I, the "Shack Sketches," is a series of parables about meadow mice, pasque flowers, chickadees, and pines above the snow. Part II, "Sketches Here and There," is a series of homilies about the untoward ecological impact of the industrialization of agriculture, wetland draining, and wilderness desecration. Perhaps most transparently biblical in intent, the second most famous essay in *Sand County*, "Thinking Like a Mountain," recalls an occasion, a quarter-century prior, when Leopold and his companions, working as forest rangers in the Southwest, murdered a mother wolf and wounded one of her pups. The moment is presented as a road-to-Damascus epiphany in which the wolf fixes Leopold (1949, 130) with her fading gaze—"a fierce green fire dying in her eyes"—and, in effect, mutely asks: why persecutest thou me? Finally, the book ends with "The Land Ethic," a sermon on the moral obligations of people to land.

Trained at the Yale Forest School, Leopold eventually became an autodidact wildlife ecologist, so he grounded the land ethic in his principal intellectual heritage: ecology and Darwinian evolutionary biology (or more exactly, Darwinian evolutionary psychology). Leopold's land ethic has become the environmental ethic of choice among his fellow conservation biologists (Groom et al. 2006)—a field that Leopold also prophetically pioneered (Meine et al. 2006)—and other environmental and conservation professionals working in the field. This is doubtless because conservation biologists and other practicing conservationists and environmentalists feel a professional and intellectual affinity with Leopold, as well as because of the accessible and compelling way that he presents the land ethic. No subsequently developed environmental ethic has gained such wide currency and fond allegiance. However, resting something enduring—and one would surely hope that an environmental ethic would endure—on scientific foundations is risky, because science is dynamic; it is self-correcting and thus ever changing. After 1949, when Leopold composed the land ethic, ecology underwent a major paradigm shift and Darwinian evolutionary biology (and psychology) became *neo*-Darwinian—and thereby putatively more rigorous and certainly more reductive. These changes in its scientific foundations are not fatal to the land ethic, but—if the land ethic is to continue to be relevant—its conceptual foundations must be reconstructed and its precepts revised accordingly.

In revised form, the land ethic remains applicable to the environmental concerns that it was conceived to address, but its applicability to other concerns is limited in two significant ways.

It is, after all, the *land* ethic. What about aquatic biotic communities and ecosystems? Leopold does state that "[t]he land ethic simply enlarges the boundaries of the land community to include soils, *waters*, plants, and animals" (1949, 204, emphasis added). But by "waters," it is clear that Leopold has freshwater streams, rivers, and lakes in mind, for such waters—with soils, plants, and animals—constitute "collectively: the land." More specifically, then, what about *marine* biotic communities and ecosystems? Nearly three-fourths of the Earth's surface is ocean. Can the land ethic be expanded to embrace seascapes as well as landscapes? In different essays published at different times and in different venues, I have defended mutually contradictory answers to that question (Callicott 1992, Callicott & Back 2008).

The land ethic is also limited by the spatial and temporal parameters to which it is scaled. It is spatially scaled to biotic communities, ecosystems, and landscapes; and its temporal scale is calibrated in decades. The spatial and temporal scales of Leopold's land-ethical thinking are clearly revealed in "Thinking Like a Mountain." There, he writes, "[F]or while a buck pulled down by wolves can be replaced in two or three years, a *range* pulled down by too many deer may fail of replacement in as many *decades*" (Leopold 1949, 132, emphasis added). For most of the twentieth century, such scales seemed large and long in comparison with ward politics and election cycles and commercial activities and economic cycles. Public concern, however, about global climate change began to be expressed in the last quarter of the twentieth century and has become increasingly acute and urgent as the twenty-first century unfolds. And the spatial scale of global climate change is planetary, while the temporal scale of the anticipated calamitous effects of global climate change—such as sea-level rise and desertification—is calibrated in centuries and millennia.

Wherever the truth may lie regarding the possibility of an oxymoronic marine land ethic, it is abundantly clear that the land ethic, even in revised form, cannot simply be scaled up to address the environmental-ethical challenge of global climate change. As noted, the scientific foundations of the land ethic are ecology and Darwinian evolutionary biology and psychology, while the sciences most centrally contributing to our understanding of global climate change are thermodynamics, geochemistry, evolutionary biology, and biological systems theory. Integrated into one interdisciplinary whole, those sciences constitute Gaian science. And the existence of a biospheric community—of which we can regard ourselves as plain members and citizens—is even more problematic than the existence of biotic communities. While the spatial dimensions of such a putative community are clear enough, how can we clearly define its temporal dimensions?

Given Aldo Leopold's cachet in environmental ethics, both among academic theorists and professional practitioners, the irrelevance of the land ethic to concern about global climate change is a pity. Fortunately, however, Leopold once more lives up to his posthumous reputation as a prophet—and, in this instance, quite amazingly. For back in 1923, he sketched, ever so sparingly, the Gaia hypothesis and a corresponding Earth—as distinct from a land—ethic. The Gaia concept coalesced in the 1970s. However, some of the leading exponents of Gaia theory, Lynn Margulis notable among them, attribute the earliest expression of Gaian thinking to Vladimir Vernadsky (1998). Vernadsky's proto-Gaian book, *The Biosphere*, was published in Russia in 1926 (Lazcano et al. 1998). While I do not suggest that the honor of being the first to broach the Gaia hypothesis should go to Leopold instead of Vernadsky—because of the faintness of Leopold's early sketch, because it remained unpublished for more than fifty years, and because his scientific and ethical thinking soon turned in a different direction—Leopold did commit his Gaia notions to writing, however briefly, three years before Vernadsky completed and published his. Of course, Vernadsky was no more aware of Leopold than was Leopold of Vernadsky, so there is no question of any influence of the thinking of the one on that of the other.

The Conceptual Foundations of the Leopold Earth Ethic

The paper in which Leopold faintly sketches the Gaia hypothesis and an associated Earth Ethic is innocuously titled "Some Fundamentals of Conservation in the Southwest." The typescript is dated 1923 and was never published during Leopold's lifetime. It eventually appeared in the first volume of *Environmental Ethics* (the journal)—the same year, coincidentally, in which James Lovelock's *Gaia: A New Look at Life on Earth* was published.[3] "Some Fundamentals" is divided into three sections, the first of which classifies the resources of the Southwest as "minerals," "organic," "climatic," "historic," and "geographic." Leopold then expressly confines his discussion to the first two, and of these, only one paragraph is devoted to minerals. As to organic resources—farms, forests, and ranges—Leopold focuses on their water and soil underpinnings. And finally his focus narrows to the alarming erosion of the region's soil—which he correctly attributed to overgrazing and fire suppression—as the ultimate concern. Soil erosion negatively affects water resources by silting reservoirs and widening formerly narrow and clear streams, making them turbid and "flashy." Leopold concludes the second section of "Some Fundamentals" with these words: "Erosion eats into our hills like a contagion, and floods bring down the loosened soil upon our valleys like a scourge" (1979, 138). This, by the way, is a good example of Leopold's early experimentation with biblical phrasing and cadence.

The third section of "Some Fundamentals" is titled "Conservation as a Moral Issue." To the philosopher's eye, Leopold appears to suggest three distinct ethical modes for addressing the moral issue that conservation represents: classical virtue ethics, consequentialistic duties to future generations, and Kantian deontological respect for the Earth per se as a living being.

> Who cannot feel the moral scorn and contempt for poor craftsmanship in the voice of Ezekiel when he asks: "Seemeth it a small thing unto you to have fed upon good pasture, but ye must tread down with your feet the residue of the pasture? And to have drunk of the clear waters, but ye must foul the residue with your feet?" ... Ezekiel seems to scorn waste, pollution, and unnecessary damage as something unworthy—as something damaging not only to the waster, but to the *self-respect* of the craft and the society of which he is a member (Leopold 1979, 138-139, emphasis added).[4]

Plato, Aristotle, and their Greek contemporaries conceived of ethics in terms of virtue, which is centrally about *self* respect. Concern or respect for others is only a benign side effect of personal *aretê*. In addition to reflecting badly on the waster himself, waste, pollution, and unnecessary damage also reflect badly on the polluter's society and the damager's craft—farming, ranching, and forestry in this case—notes Leopold, quite consistently with Greek thinking in this regard. Technically put, virtue ethics is not exclusively individualistic but also has at least two holistic dimensions: the professional self-respect of the guild as such and the self-respect of the collective society per se.

> We might even draw from his [Ezekiel's] words a broader concept—that the privilege of possessing the earth entails the responsibility of passing it on, the better for our use, not only to immediate posterity, but to the Unknown Future, the nature of which it is not given us to know (Leopold 1979, 139).

Here Leopold does suggest that in addition to self-respect—individually, professionally, and socially—we should also be concerned about the effect of waste, pollution, and damage on the welfare of other humans who will follow us. This adumbrates an anthropocentric utilitarian mode of ethical thought. Here, too, we have both an individualistic and holistic aspect. Immediate posterity is an aggregate of identifiable individuals: our existing children and grandchildren and those who will exist in our own lifetimes or in the lifetimes of those whom we know and care about. Distant future generations are conceivable only collectively, not as a set of identifiable individuals. Perhaps to bring this holistic aspect out, Leopold refers to them as the *Unknown* Future, and goes on, redundantly, to stress our irremediable ignorance regarding just who they might turn out to be.

> It is possible that Ezekiel respected the soil, not only as a craftsman respects his material, but as a moral being respects a living thing (Leopold 1979, 139).

Having suggested three distinct ethical modes for conceiving of conservation as a moral issue all in one paragraph, Leopold devotes the rest of the essay to elaborating upon the third: respect for the soil—which immediately becomes "the earth"—as a living being. We cannot coherently imagine that our waste, pollution, and damage will adversely affect the *welfare* of the Earth—its happiness or, as contemporary utilitarians would express it, the Earth's degree of "preference satisfaction." But we can coherently imagine that waste, pollution, and damage violate our duty to *respect* the Earth—as a moral being respects a living thing. Respect for beings with intrinsic value is a hallmark of Kant's deontological (or duty-oriented) ethic. Kant, however, was a militant ratiocentrist, insisting that being rational is a sine qua non of an entity's intrinsic value, dignity, and respect-worthiness. But—excepting, perhaps, God and the heavenly host—the only rational beings that Kant recognized to actually exist were human beings. Hence, Kant's ethics is also, *in effect*, militantly anthropocentric.

Leopold takes as his first task persuading his audience that the Earth is indeed a living being, and it is by that effort that Leopold anticipates the Gaia hypothesis. His second task is to undermine his audience's knee-jerk anthropocentrism, a legacy of Western religion and philosophy. I here review only the first.

Leopold (1979, 139) presciently notes that "[t]he very words *living thing* have an inherited and arbitrary meaning derived not from reality, but from human perceptions of human affairs." As Gaian theorists have indicated, life may better be understood in terms of its recursive formal properties—its autopoiesis and autonomy—rather than in the classical biological terms of morphology and reproduction. Leopold goes on to quote and paraphrase *Tertium Organum*—by another Russian, coincidentally, P. D. Ouspensky (1922). Its title may suggest that the book is all about superorganisms, or third-order organisms like Gaia. First-order organisms, one may suppose, are single-celled; second-order organisms are multi-celled; and third-order organisms are to multi-celled organisms as multi-celled organisms are to single-celled organisms. But that's not what *Tertium Organum* is about—not at all—nor is that the meaning of the title. Aristotle's corpus of treatises on logic is called the *Organon*. Francis Bacon wrote a *Novum Organum*, published in 1620, a new epistemology for the nascent modern science. Ouspensky hubristically intended his *Tertium Organum* to supersede Aristotle's *Organon* and Bacon's *Novum Organum*. And despite its sensational popularity when Leopold got hold of it, Ouspensky's book is really as vacuous as it is pompous and pretentious. Nor can I find any evidence that Ouspensky was acquainted with the work of Vernadsky or vice versa. So what Leopold goes on to attribute to Ouspensky are really his own speculations:

> [I]t is at least not impossible to regard the earth's parts—soils, mountains, rivers, atmosphere, etc.—as organs or parts of organs, of a coordinated whole, each part with a definite function. And, if we could see this whole, as a whole, through a great period of time, we might perceive not only organs with coordinated functions, but possibly also that process of consumption and replacement which in biology we call metabolism, or growth. In such a case we would have all the visible attributes of a living thing, which we do not now realize to be such because it is too big, and its life processes too slow (1979, 139).

That this is indeed an early—perhaps the very first—instance of Gaian thinking is suggested by Leopold's attribution of a metabolism to the Earth as a whole. Metabolism is a defining characteristic of living organisms—which are in a far-from-equilibrium thermodynamic state, open to energy flows but closed in regard to their own processes. Earth's putative "organs with coordinated functions" represent its operational closure, while its putative "metabolism" is constituted by its openness to solar energy, gravitational influences from the sun and moon, and ambient cosmic materials.

Leopold goes on to suggest that if the Earth has both organs with coordinated functions and a metabolism, that "there would also follow that invisible attribute—a soul or consciousness—which … many philosophers of all ages ascribe to all living things and aggregations thereof, including the 'dead' earth" (ibid.). This may immediately raise suspicions that Leopold's Gaian speculations are as fraught with teleological tendencies as the early Lovelock's—that Gaia was consciously planning and directing her own autopoiesis and evolution (Bormann 1981). But as each of us should well know simply by introspection, being conscious, as we are, by no means implies that we plan and direct our own ontogeny, physiology, and metabolism. Gaia's consciousness, like our own, might well be emergent and epiphenomenal and oriented toward perception and reaction to environmental changes—such as changes in the amplitude of solar radiation—rather than toward its own internal organization and processes. Leopold concludes his brief propaedeutic to a future Gaia hypothesis with a coda, fusing science, ethics, and poetry:

> Possibly in our intuitive perceptions, which may be truer than our science and less impeded by words than our philosophies, we realize the indivisibility of the earth—its soil, mountains, rivers, forests, climate, plants, and animals, and respect it collectively not only as a useful servant but as a living being, vastly less alive than ourselves in time and space—a being that was old when the morning stars sang together, and, when the last of us has been gathered unto our fathers, will still be young (1979, 140).

An Earth Ethic Rising

So what shape should an Earth ethic assume? As noted, Leopold hints that it should be essentially Kantian, based on deontological respect. About this possibility, however, he seems unduly sanguine: "Philosophy, then, suggests one reason why we cannot destroy the earth with moral impunity; namely, that the 'dead' earth is an organism possessing a certain kind and degree of life, *which we intuitively respect as such*" (Leopold 1979, 140). But do we? The indifference most people seem to show to what Holmes Rolston (1994) calls "super-killing"—anthropogenic species extinction—let alone ordinary killing (anthropogenic destruction of individual organisms) makes one wonder if we really do intuitively respect life as such. Maybe that's why Leopold immediately turns to ridiculing Western anthropocentrism.

Perhaps an adequate and effective Earth ethic should be pluralistic. And indeed, Leopold seems to agree. While he devotes most of "Conservation as a Moral Issue" to his call for respecting the Earth as a living being, he begins, as noted, by invoking two other ethical modes: individual, professional, and social virtue; and concern for the welfare of future generations, both immediate posterity and the Unknown Future. I would add a fourth. Global climate change will not be felt equally by all members of the present generation. People living in low-lying deltas, such as that of the Ganges, and on oceanic atolls will suffer disproportionately, first from more incidental flooding and salt-water intrusion, and finally eviction as sea level rises and storm surges intensify (IPCC 2007). People living in almost all tropical and temperate coastal areas will suffer from the increased frequency and intensity of cyclones (IPCC 2007). Indigenous peoples of the Arctic will find their environments and thus their traditional cultures more radically altered by warming than peoples living at lower latitudes (IPCC 2007). In respect to all these untoward phenomena and many others, the more affluent will be better able to cope than the less affluent. Adding insult to injury, those least responsible for generating greenhouse gases, living in nonindustrialized societies, will suffer, on average, more than those living in industrialized societies, who are most responsible. Such injustices demand redress, and that is what environmental-justice ethics is all about (Bauer 2006).

The spatial scale of global climate change requires a profound shift in our moral ontology. Ethics, as we know it, has assumed an atomistic ontology—moral agents and patients are thought to be individuals. But as these considerations of environmental justice suggest, the only effective moral agents and patients at the global scale are societies, not individuals. What compensation, we are asking, in effect, do less-at-risk-but-more-to-blame industrialized *societies* owe more-at-risk-but-less-to-blame non-industrialized *societies*? No environmental-justice ethicist insists that affluent American or European individuals should write personal checks to individual impecunious Micronesians in order to compensate them for sea-level rise and to enable them to cope with that.

The temporal scale of global climate change makes the necessity for a shift from an individualistic to a socio-cultural moral ontology even more obvious. We used to marvel at the temporal horizons of moral deliberations among the Iroquois—who considered the effect of present choices out seven generations (Erdrich 1996). But, do the math. Assume a twenty-five-year interval for each generation—a reasonable age-span between human parents and their offspring: 7 x 25 = 175 + 75 (the life span of the seventh generation) = 250 years.

The serious adverse effects of global climate change, however, are predicted to *begin* to be registered only by mid-century, and not to kick into full tilt for several centuries hence, and then to last for millennia (IPCC 2007; Solomon et al. 2009). And the lag time for realizing the effects of remedial actions we take now will occur on the same temporal scale—calibrated in centuries and millennia. Does one care—can one care—about the welfare of *individual* human beings living in the twenty-third century or the twenty-fourth? What about those living in the thirtieth century or the fortieth?

To that question, one may at least reply that we do in fact care, and care deeply, about individual human beings who lived 2500 years ago, even 3000 years ago. Who can read the *Apology* and not share Plato's bitter condemnation of Anytos, Lycon, and Meletus and the kangaroo court of Athens that unjustly convicted Socrates of impiety and corrupting the youth and sentenced him to death? Who can read the *Odyssey* and not seethe with the same righteous indignation that consumed Telemachus because of the outrageous behavior of the profligate and calumnious suitors? But that is because we know Socrates and Telemachus as individuals. We care far less about the Unknown Past beyond the pale of history or, for that matter, about the little known individuals who suffered various slings and arrows of outrageous fortune during the European Dark Ages. Of course we care and care deeply about our children and grandchildren, whom we know personally and love deeply, and about the future world that they will inherit from us. And because I care about my grandson, I care about what he cares about, and one day he will care about his own grandchildren. Thus, our moral sentiments can and do reach out and touch future individuals about a century and a half hence. But I confess I cannot summon up much concern, individually, for the great grandchildren of my grandson, or for their children or grandchildren. As Leopold points out, beyond immediate posterity, lies the "*Unknown* Future." Can we muster up the motivation to make sacrifices now that will benefit anonymous individual people living one, two, or three thousand years hence?

Conclusion

At the very beginning of "The Land Ethic," Leopold points out that Western *civilization* goes back three thousand years to the time of Homer. If our moral ontology is socio-culturally scaled, as opposed to individually scaled, and the survival and continuity of human civilization is what we are concerned about, not the welfare of unknown future individuals severally, what would the math look like then? We conventionally divide the history of Western civilization into three periods—ancient, medieval, and modern—each roughly a millennium in duration. So seven socio-cultural generations would come out to be seven thousand years.

So much for a moral-*patient* ontology scaled to the temporal parameters of global climate change; what about a moral-*agent* ontology? I was appalled by what I saw at the end of Al Gore's otherwise excellent documentary, *An Inconvenient Truth*: a list of things that each of us, *individually and voluntarily*, can do to reduce our carbon emissions. I myself do many of those things: replace halogen light bulbs with compact fluorescents; make my home-to-office-and-back commute by bicycle; etc. But I live in Denton, Texas—not Ashland, Oregon or Boulder, Colorado. So I am painfully aware that my individual efforts to lessen the size and impress of my own personal carbon footprint are swamped by the recalcitrance of the overwhelming majority of my fellow citizens. Many of them have never heard of global climate change. Many of those who have are convinced that it is a hoax cooked up by self-righteous environmentalist elites who cannot stand to see common people have their mechanized fun. And many of those who think that it is for real welcome it as a sign that the End Times are near, the horrors of which they will be spared by the Rapture. It will not suffice, therefore, simply to encourage people *individually and voluntarily* to build green and drive hybrid. But what is worse is the implication that that is all we can do about it, that the ultimate responsibility for dampening the adverse effects of global climate change devolves to each of us as individuals.

Taking individual action is, indeed, important, but not because of the aggregate effects of individual action. Rather, individual action can result in the self-organization of collective phenomena. A relevant example is the Step It Up National Day of Climate Action in the United States in which thousands of people remonstrated the Bush Administration on April 14 and again on November 3, 2007, for its inaction on the dire threat of global climate change. Such spontaneous self-organization is facilitated, if not entirely made possible by, the Internet. And it leads, hopefully, to the only kind of action that is effective: collective action through collective institutions. And what are those institutions? Among the most powerful are national governments and international institutions. The only hope we have to temper global climate change is a collective socio-cultural response in the form of policy, regulation, treaty, and law. What is required, in the closing words of "Tragedy of the Commons," is "mutual coercion mutually agreed upon" (Hardin 1968, 1248).

Notes

1. The prophet epithet appears traceable to Roberts Mann (1954); it was picked up by Ernest Swift (1961); Roderick Nash (1967) institutionalized it in *Wilderness and the American Mind*, titling one chapter in that classic, "Aldo Leopold: Prophet."
2. The world's first college course titled "environmental ethics" was offered in 1971 at the University of Wisconsin-Stevens Point. The first articles by academic philosophers were published two years later (Routley 1973; Naess 1973).
3. "Some Fundamentals of Conservation in the Southwest," was slightly edited and reprinted (Flader and Callicott, 1989). One noteworthy change was made. The typescript spelled Ouspensky's name as "Onpensky." But there is no doubt that Leopold referred to Ouspensky because he accurately quotes from *Tertium Organum* verbatim. The published article in *Environmental Ethics* did not correct the spelling of Ouspensky's name. Feeling that the error was introduced by Leopold's secretary as she typed up his manuscript, I corrected her error for republication in *RMG*. Leopold composed in pencil in a tight cursive script and so, quite excusably, his secretary read his "us" as an "n."
4. Leopold quotes from Richard G. Moulton, *Modern Reader's Bible* (New York: Macmillan, 1907). The editor and translator of Leopold's bible, Richard G. Moulton, was also author of *The Literary Study of the Bible: An Account of the Leading Forms of Literature Represented in the Sacred Writings* (Lexington, Mass.: D. C. Heath, 1899). This fact would seem to confirm that Leopold's interest in the bible was less devotional than rhetorical.

References

Bauer, J. (2006) *Forging Environmentalism: Justice, Livelihood, and Contested Environments*. New York: M.E. Sharpe.

Bormann, F. H. (1981) "The Gaia Hypothesis," (review of *Gaia: A New Look at Life on Earth* by James Lovelock). *Ecology*, 62: 502.

Callicott, J. B. (1992) "Principal Traditions in American Environmental Ethics: A Survey of Moral Values for Framing an American Ocean Policy," *Ocean and Shoreline Management*, 17: 299-308.

Callicott, J. B. and E. Back (2008) "The Conceptual Foundations of Rachel Carson's Sea Ethic," in Lisa Sideris and Kathleen Dean Moore (eds.), *The Philosophy of Rachel Carson*. Albany: State Universiy of New York Press, pp. 000-000.

Carson, R. (1962) *Silent Spring*. Boston: Houghon Mifflin.

Erdrich, L. (1996) "Read Their Lips! Three Novel Ideas for a Clinton Speech." *Washington Post*, June 23: C-1.

Flader, S.L. and Callicott, J.B. ed., (1989) *The River of the Mother of God and Other Essays by Aldo Leopold*. Madison: University of Wisconsin Press.

Groom, M. J., G. K. Meffe and C. R. Carroll (2006) *Principles of Conservation Biology*. Sunderland, MA: Sinauer Associates.

Hardin, G. (1968) "Tragedy of the Commons," *Science*, 162: 1243-1248.

Intergovernmental Panel on Climate Change (IPCC) (2007) *Fourth Assessment Report*.

Lazcano, D. Suzuki, C. Tickell, M. Walter and P. Wesybroek (1998) "Forward to the English Translation," in V. I. Vernadsky, *The Biosphere*. New York: Copernicus, pp. 14-19.

Leopold, A. (1949) *A Sand County Almanac and Sketches Here and There*. New York: Oxford University Press.

Leopold, A. (1979) "Some Fundamentals of Conservation in the Southwest." *Environmental Ethics*, 1: 131-141.
Mann, R. (1954) "Aldo Leopold: Priest and Prophet." *American Forests*, 60/8 (August): 23, 42-43.
Naess, A. (1973) "The Shallow and the Deep, Long-range Ecology Movements: A Summary." *Inquiry*, 16: 95-100.
Nash, R. *Wilderness and the American Mind.* New Haven, CT: Yale University Press.
Ouspensky, P. D. (1922) *Tertium Organum: The Third Cannon of Thought; A Key to the Enigmas of the World.* Nicholas Bessaraboff and Claude Bragdon, trans. New York: Knopf.
Rolston, H. (1994) *Conserving Natural Value*. New York: Columbia University Press.
Routley, R. (1973) "Is There a Need for a New, an Environmental Ethic?," in Bulgarian Organizing Committee (ed.), *Proceedings of the Fifteenth World Congress of Philosophy*. Varna: Sophia Press, pp. 205-210.
Solomon, S., P. G-K. Pattner, R. Knutti and P. Friedlingstein (2009) "Irreversible Climate Change Due to Carbon Dioxide Emissions." *Proceedings of the National Academy of Sciences*, doi: 10.1073/pnas 0812721106.
Stegner, W. (1989) "The Legacy of Aldo Leopold," in J. Baird Callicott (ed.), *Companion to* A Sand County Almanac*: Interpretive and Critical Essays*. Madison: University of Wisconsin Press, pp. 233-245.
Swift, E. (1961) "Aldo Leopold: Wisconsin's Conservation Prophet." *Wisconsin Tales and Trails*, 2/2 (September): 2-5.
Tallmadge, J. (1989) "Anatomy of a Classic," in J. Baird Callicott (ed.), *Companion to* A Sand County Almanac*: Interpretive and Critical Essays*. Madison: University of Wisconsin Press, pp. 110-127.
Udall, S. L. (1963) *The Quiet Crisis*. New York: Holt, Rinehart and Winston.
Vernadsky, V. I. (1998) *The Biosphere*. New York: Copernicus.

2

Climate Change, Environmental Ethics, and Biocentrism

Robin Attfield

When environmental philosophy began in the early seventies, global warming was unknown. Several early participants, concerned about the loss of species and habitats, rejected anthropocentrism, preferring biocentric or ecocentric theories. Anthropocentrism then made a comeback, and many prominent environmental philosophers now espouse it. However, the discovery of systemic global environmental problems (including ozone-depletion as well as global warming) raises the issue of whether an anthropocentric ethic can tackle them adequately. So does the issue of cruelty to animals, the ethics of which seem to require a non-anthropocentric theory, albeit possibly one no more radical than sentientism. Sentientism, however, is open to a range of objections, including Val Plumwood's charge of desensitized rationalism, which already suggests that biocentrism is preferable. Tackling global warming, for its part, requires principles of equity, which may well warrant authorizing equal entitlements to emit greenhouse gases to all living human beings, up to an agreed ceiling, which would itself be lowered in the course of time (Contraction and Convergence). Issues that arise (and that are now discussed) concern the compatibility of biocentrism with Contraction and Convergence, and also whether biocentrism makes a difference to the solution that is needed. It is concluded that Contraction and Convergence can be reconciled with biocentrism, and that biocentrism strengthens the case for fixing the ceiling for greenhouse gas emissions below the actual current level. Ecocentrism is considered as an alternative ethic, but reasons are given for rejecting it. Biocentrism, it is concluded, is intrinsically superior as an ethic, and significantly strengthens the case for strong international action on climate change.

Anthropocentrism, Biocentrism, and Environmental Philosophy

Global warming had not been heard of, let alone entered the conventional wisdom, in the early days of environmental philosophy in the

1970s. When philosophers such as Holmes Rolston, Richard and Val Routley, and Arne Naess challenged anthropocentrism and pioneered biocentric or ecocentric value-theories, the practical issues were deforestation, pesticides, desertification, loss of habitats, and sometimes, though more controversially, population (Rolston 1975, Routley 1973, Naess 1973, Routley & Routley 1980,[1] Routley & Routley 1975). A few years later, in writing *The Ethics of Environmental Concern* (1983, 1991), my position was one of broadening biocentrism, qualifying concerns about population, introducing some of Derek Parfit's themes concerning obligations to future generations, and supplying nuanced support for John Passmore's view that Western ethical traditions were capable of sustaining a nature-friendly approach, although I insisted that the promising strands were ones that rejected anthropocentrism in favor of biocentrism (Attfield 1983, 1991; Parfit 1984; Passmore 1974). In particular, I was advocating a broadened version of consequentialism, biocentric consequentialism, as I have done since (Attfield 1994, Attfield 1995, Attfield 1999, Attfield 2003).

Subsequently, anthropocentrism made a comeback. Bryan Norton argued that anthropocentric and non-anthropocentric stances sustained exactly the same policies, and that his weak anthropocentrism, which turned on human concerns as well as human interests, was a sufficient environmental ethic (Norton 1991). Variants on this stance were supported by David Cooper, Eugene Hargrove, John Benson, and John O'Neill (Cooper 1995, Hargrove 1989, Benson 2000, O'Neill 1993, O'Neill, Holland & Light 2008), the latter adducing an Aristotelian virtue-ethics approach, which claimed that preserving natural creatures was so clearly in the human interest (since our lives would be warped without them) that this was the underlying reason for preservation. (I have replied to this approach in a recent book, but it seems to be gaining in popularity. While I agree that we need an Aristotelian understanding of human well-being, this seems enormously insufficient to underpin the preservation of species such as those that have not yet been discovered and about which no one has had the chance to care (Attfield 2003, 69-72).) This phase of environmental ethics has generated a salutary increase of emphasis on responsibilities to our successors, boosted by the writings of Hans Jonas, who stresses the enlarged impact of human actions, and whose writings have resonated in continental philosophy circles as well as analytic ones. Yet it should be remembered that Jonas too is a critic of anthropocentrism (Jonas 1984).

So too was Val Plumwood, previously Val Routley, who in addition, argued for a fuller recognition of responsibilities towards the future (particularly in matters of nuclear energy generation) (Routley & Routley 1978), and for a broader rationality that would take account (in a feminist manner) of our embodiedness and of our emotions (Plumwood 1991). This kind of approach, to my mind, rightly contested the human-centered aspects of Aristotle's metaphysic, and recognized that there is value (and not only instrumental value) in the realm beyond humanity, or, in other words, that there are reasons for preserving all manner of living creatures, not for our sakes, but for theirs.

Anthropocentrism, Sentientism, and Global Warming

During this period, the phenomenon of global warming, and of its largely anthropogenic causes, became increasingly acknowledged, not least through successive reports of the Intergovernmental Panel on Climate Change (IPCC) from the 1990s onwards. This recognition included awareness that at least some environmental problems are systemic global problems, not just present in many places (as traffic congestion is), but reflecting interlocking systems affected everywhere by human actions in each and every longitude and latitude. Global warming is not the only such problem; erosion of the ozone layer through the use of propellants and refrigerants such as CFCs and HCFCs was found to be another, but was rapidly attended to through global agreements to ban the production and exporting of these substances, and (just in time) to finance the technology transfer needed for Third World countries to produce substitutes (Tudge 1995, Yearley 1996). Such problems showed that human action was endangering the entire biosphere, and for all future ages. Indeed from this stage onwards, philosophers such as Finn Arler and Henry Shue began reflecting on the bearing of global or international equity for climate-change-related regimes in tackling the problems of climate change (Arler 1995, Shue 1995).

Can an anthropocentric ethic cope with systemic global problems of climate change? To address this question, the first response is that it cannot cope with much more familiar issues, such as the wrongness of cruelty to animals. It is not plausible that the only reason why such cruelty is wrong is human welfare, whether because cruelty to animals corrupts the character of the agent, or because harm to animals damages human property. This is because such cruelty would still be wrong even if these undesirable impacts failed to happen. Thus unqualified anthropocentrism abjectly fails in any case, whether in a weak or a strong form, and whether Aristotelian or not. But if it is modified into an ethic based on avoidance of suffering for sentient creatures, plus the promotion of their well-being, or sentientism, the kind of stance adopted by Janna Thompson and by Peter Singer (Thompson 1990, Singer 1993), then the resulting theory is not so obviously defective.

This changes the question as to whether sentientism can cope with problems like climate change. While it may seem capable of supplying reasons for most land-based ecosystems, it is less clear that it does so for those of the deep oceans, the denizens of which may or may not be sentient, but which pursue their own good regardless. Besides, at the level of theory it seems unconvincing. For it is not only conscious states that are valuable; others, such as the health of creatures, whether sentient, conscious, or neither, can hardly be denied value as well. Would not a planet with self-reproducing and self-repairing life, capable perhaps of photosynthesis and respiration but not of sentience, be worth saving, if this were the only life we could save from a dying universe (Scherer 1983)? Besides, the view that only sentient creatures should be the objects of our direct concern seems open to Val Routley's kind of criticism of disembodied rationalism; it suggests that we should reason by analogy from human interests to animal ones, but cannot or at least should not recognize any kinship with other creatures of the evolutionary tree, ones that have solved all manner of problems of survival but may not have done so through adaptations involving feelings. Thus the sphere where human action can cause benefit or harm extends beyond the sphere of sentience, as far as the limits of life. This also means that it extends further into the ocean depths, and probably further into the future, than the implications of either anthropocentrism or sentientism would suggest. And this in turn, as will be seen, has implications for acceptable levels of greenhouse gases in the atmosphere.

Admittedly, the impacts of human action and inaction are often uncertain, and this is sometimes thought to diminish our responsibilities. However, to inflict risks of significant harm on others who have not chosen them is standardly treated in a variety of legal systems not only as neglectful but as positively reckless, and on both counts as criminal behavior. This applies in traffic law, and in laws governing the need for safety precautions where there are risks of possible harm through noxious substances or dangerous processes. Just the same applies, I suggest, where anthropogenic changes to weather systems foreseeably introduce new or heightened risks of extreme weather events (floods, droughts or firestorms), whether for humans or other creatures, usually for victims who have no choice in the matter and have had little or no role in causing the increased hazards to which they are subjected. Donald Brown and his co-authors of the "White Paper on the Ethical Implications of Climate Change" have rightly stressed this analogy between moral attitudes to inter-human recklessness and the attitudes that are in place to climatic recklessness, whether the victims belong to different nations, generations, or species (Brown et al. 2006, 26-27). But this also means that failure to act when action is required to restrict such risks is to be condemned.

Moreover, principles are also needed for the allocation and restriction of carbon emissions and the emission of carbon-equivalent gases. Whatever an acceptable level of greenhouse gases may be, it is difficult, if not impossible, to justify any human being having a greater entitlement than any other to emit these gases (the principle of equality). The fact that one's ancestors emitted more such gases before the theory of anthropogenic global warming came to light in the 1980s fails in my view to justify reducing this entitlement, since those emissions were discharged in ignorance and not known to take place at others' expense. Equally, the fact that the status quo and the current world economic system implicate much greater emissions for developed countries than this principle would recognize fails to justify these countries or their peoples retaining this differential, or being allowed differential entitlements. So, if countries are allowed to act and to exercise responsibilities on behalf of their populations, then the entitlements of countries should be proportional to their populations (as calculated at some agreed date). An international regime should, if so, be introduced to give effect to such entitlements, a regime that would authorize countries not using their full entitlement to trade the unused component with countries wishing to exceed their entitlement. This would clearly be a redistributive system, even if the acceptable total were to be steadily reduced to stabilize total emissions. This is the system of Contraction and Convergence, proposed by Aubrey Meyer, and variously defended by Peter Singer, Dale Jamieson, and myself (Meyer 2005; Singer 2002; Jamieson 2005; Attfield 2003, 179-181).

Contraction, Convergence, and Equity

How well does such a system accord with ethical theory, particularly if anthropocentrism and sentientism are to be rejected in favor of a more biocentric approach? I want to tackle here some issues of two different kinds. First, there are issues of the consistency of the Contraction and

Convergence approach with biocentrism, and its recognition of the importance of making proper provisions for nonhuman species. Secondly, there are issues relating to the difference made by biocentrism to what we should aim at in a global regime to cope with climate change and related issues.

On the face of it, Contraction and Convergence could be accused of anthropocentrism, since the entitlements that it recognizes are for human beings and for them alone. This might almost seem like a human takeover of the atmosphere's absorptive capacities. Even though humans depend on a whole range of ecosystems, the functioning of which would have to be provided for, this recognition still seems to derive from human interests alone, and not to embody the least concern for other species.

However, calculations of emission entitlements would need to take into account the normal functioning of ecosystems whether they benefit humanity or not. Thus the methane buried in temperate wetlands and in tundra has to be allowed for, since its emission is largely beyond human control. Admittedly, the mitigating of greenhouse gas emissions requires not exacerbating these emissions, but their lack of benefit to humanity does not mean that they can be disregarded, or that their contribution to the proportion of carbon-equivalent gases in the atmosphere can or should be forgotten. Much the same applies to the emissions both of oxygen (welcome) and of carbon dioxide (less welcome) from tropical forests and from oceanic vegetation. These ecological processes are part of the background to issues about the shape of global climate agreements, and Contraction and Convergence has no tendency not to take them into account.

Much the same should be said about the emissions of wild animals, whether plentiful ones like bees and ants or rare ones like tigers and pandas. Some of these species are vital for human interests; among the species just mentioned, bees are the clearest example. However, even the kinds that are not, such as perhaps snakes and spiders, must be recognized as having their own patterns of ingestion and excretion, just like the trees, the plankton, and the seaweeds discussed implicitly in connection with forest and ocean ecosystems. Any attempt to appropriate or seize their ecological niches would be both arrogant and disastrous, except where, as pests, they need to be controlled to allow human food to be grown and stored. Biocentric theorists can welcome these necessities, where anthropocentric ones may regard them with resignation, but both kinds of theorist are free to support Contraction and Convergence in at least some of its varieties, for Contraction and Convergence has no tendency to colonize the entire surface of the planet in the cause of policing emissions. Besides, there are strong grounds for preserving wild species, and supporters of Contraction and Convergence, whether anthropocentric or biocentric, have no need to disregard them.

Where domestic animals are concerned, the situation is different, since their numbers, and to some degree, their kinds, are subject to human control. Accordingly, the emissions of such creatures are to be regarded as part of the tally of human emissions. If, as it might, Contraction involves rearing fewer heads of livestock and fostering a more vegetarian diet, this possible implication would have to be carried through as part of the human responsibility. There would probably be other responsibilities to preserve the various domesticated species, if not their current populations, but such responsibilities could readily be reconciled with an agreed climate regimen. Maybe these responsibilities would focus on the good of our human successors, or maybe they would relate to possible future members of nonhuman kinds, and to their welfare. So far, then, I conclude that there is nothing objectionable about the way in which Contraction and Convergence focuses on human entitlements. There would be, if its advocates were to claim that non-bearers of these entitlements only ever carried instrumental value; but there is not the least requirement of rationality or consistency for them to say this.

Contraction, Convergence, and Biocentrism

It is time to turn to the second kind of issue mentioned earlier, and to ask what difference non-anthropocentric kinds of environmental ethics make in matters of climate regimens. So far, I have claimed that all kinds of theories of environmental ethics can support Contraction and Convergence in some form or other. But might sentientism and biocentrism make a difference as to which form is to be favored?

In principle, they must make a difference, because they supplement the human interests to be considered with the interests of billions of sentient nonhumans, and in the case of biocentrism, non-sentient creatures in their trillions. If we add to these interests the interests of future members of those species, their accumulated strength is vast. These interests would usually be added to the scales in favor of policies of mitigation, since in the absence of such policies, numerous species are at risk of extinction, many of them species with a strong prospect of survival well beyond the eventual demise of humanity, unless they are eliminated in the near future. While some creatures would doubtless benefit from the demise or decimation of humanity, the ecosystems on which most wildlife depends could well be at risk if policies of mitigation are not adopted by human agents, or adopted too feebly or too late.

Certainly there are human interests to be weighed against these policies, or which involve competition for resources. But in many cases, there would be ways of combining the policies of mitigation and adaptation with the policies that these interests support, such as policies of development, which could be combined with climate change policies, and in some cases enhanced by them. For example, during the first few decades of Contraction and Convergence, resources would flow to poor but populous countries that were not yet in a position to deploy their full emissions entitlement, and which might well decide to trade the unused component; such resources could be used both towards their own adaptation and for development.

In practice, what is at issue concerns the emissions cuts needed to prevent a two-degree (Celsius) increase in temperatures above pre-Industrial Revolution averages. Conventional policies, for example, ones tolerating 450 ppm of carbon dioxide in the atmosphere, could allow this to happen, despite purporting to prevent it. Current levels are 380 ppm of carbon dioxide, plus 60 ppm of carbon-equivalent greenhouse gases (methane, nitrous oxide, etc.) (Monbiot 2006, 15-17), but to ensure that the two-degree increase is avoided, the total level of carbon plus carbon-equivalent gases would apparently need to fall to a total of 400 ppm (Monbiot, 2006, 15-17; Brown 2002; Baer, Athanasiou & Kartha 2007). With levels continuing to rise, attaining this level is likely to involve cuts that are both early and severe.

From an anthropocentric perspective, the case for such early and severe cuts might seem less than secure, in view of losses to productivity and to the desirable attainments that productivity can support. Here, then, it is of great importance that the ethical case not be confined to the limited scope of anthropocentrism. Sentientism strengthens the case to some degree, requiring agents to heed the difference that can be made to mammals, birds, and perhaps reptiles and fish. But a much greater difference is made when the interests of the non-sentient majority of creatures are added. I am not suggesting that government negotiators are likely to be impressed by the numbers involved. Relevant considerations are likelier to be the arrogance of disregarding both current non-human life, and the future of life on earth in general. Plausibly, biocentrism would justify reducing the atmospheric concentration of greenhouse gases to below 400 ppm of carbon plus carbon-equivalent gases; it certainly indicates this much more securely than sentientism and very much more securely than anthropocentrism. (This point is made by Donald Brown (Brown 2002) in chapter 12.) All these levels are, of course, consistent with one version of Contraction and Convergence or another, but they make a very large difference as to which version is adopted, and to which forms of energy-generation, production, travel, and transport are selected.

Someone might here suggest that ecocentrism would sustain an even stronger case. Here, I beg to differ. For all the living creatures belonging to ecosystems have been included already within biocentrism, and the claim that ecosystems count for themselves independently of the creatures they sustain thus amounts to advocacy of double-counting. Certainly ecosystems are important, but their importance, I suggest, lies in the value of the creatures that they support and can continue to support, rather than in some independent value of their own. No doubt defenders of ecocentrism would claim to have a yet stronger case to present for emission reductions; my view, however, is that it is not a good case, and that environmental ethicists should appeal to biocentrism instead.

Conclusion

It is, once again, of the greatest importance that not all kinds of theories of environmental ethics support the same policies; this kind of convergence view (Norton 1991) is surely misguided. Biocentrism supports far stronger policies than anthropocentrism, however 'weak,' Aristotelian or enlightened, and considerably stronger policies than sentientism. Since it is also a more grounded theory, these stronger policies should be adopted for that reason. Hence the best way to face climate change is to appeal to a renewed environmental ethic of a non-anthropocentric, biocentric kind.

Notes

1. *A paper drafted in 1973.*

References

Arler, Finn (1995) "Justice in the Air: Energy Policy, Greenhouse Effect, and the Question of Global Justice." *Human Ecology Review*, 2, 40-61.

Attfield, Robin (1983) *The Ethics of Environmental Concern*. Oxford: Basil Blackwell and New York: Columbia University Press, 1983; 2nd edn. (1991), Athens, GA and London: University of Georgia Press.

Attfield, Robin (1994) *Environmental Philosophy: Principles and Prospects*. Aldershot, England and Brookfield, VT: Ashgate.

Attfield, Robin (1995) *Value, Obligation and Meta-Ethics*. Amsterdam and Atlanta, GA: Rodopi.

Attfield, Robin (1999) *The Ethics of the Global Environment*. Edinburgh: Edinburgh University Press and West Lafayette, IN: Purdue University Press.

Attfield, Robin (2003) *Environmental Ethics: An Overview for the Twenty-First Century*. Cambridge, England: Polity, and Malden, MA: Blackwell.

Baer, Paul, Tom Athanasiou and Sivan Kartha (2007) *The Right to Development in a Climate Constrained World: The Greenhouse Development Rights Framework*. EcoEquity: retrieved on 15 January 2009 from <www.ecoequity.org/docs/TheGDRsFramework.pdf>.

Benson, John (2000) *Environmental Ethics: An Introduction with Readings*. London and New York: London and New York, Routledge.

Brown, Donald A. (2002) *American Heat: Ethical Problems with the United States' Response to Global Warming*. Lanham, MD and Oxford: Rowman & Littlefield.

Brown, Donald A. et al. (2006) *White Paper on the Ethical Dimensions of Climate Change*. Philadelphia: Rock Ethics Institute.

Cooper, David E. (1995) "Other Species and Moral Reason," in David E. Cooper and Joy A. Palmer (eds.), *Just Environments: Intergenerational, International and Interspecies Issues*. London: Routledge, pp. 137-148.

Hargrove, Eugene C. (1989) *Foundations of Environmental Ethics*. Englewood Cliffs, N.J.: Prentice-Hall.

Jamieson, Dale (2005) "Adaptation, Mitigation and Justice," in Walter Sinnott-Armstrong and Richard B. Howarth (eds.), *Perspectives on Climate Change: Science, Economics, Politics, Ethics: Advances in the Economics of Environmental Research*. Vol. 5, 217-248.

Jonas, Hans (1984) *The Imperative of Responsibility: In Search of an Ethics for the Technological Age*. Trans. Hans Jonas and David Herr, Chicago and London: University of Chicago Press; originally published as Hans Jonas (1979) *Das Prinzip Verantwortung: Versuch einer Ethic für die technologische Zivilisation*. Insel Verlag Frankfurt am Main, and Hans Jonas (1981) *Macht oder Ohnmacht der Subjecktivität? Das Leib-Seele-Problem im Vorfeld desPrinzips Verantwortung*. Insel Verlag Frankfurt am Main.

Meyer, Aubrey (2005) *Contraction & Convergence, The Global Solution to Climate Change*. Schumacher Briefing no. 5, Totnes, Devon: Green Books.

Monbiot, George (2006) *Heat: How to Stop the Planet Burning*. London: Allen Lane.

Naess, Arne (1973) "The Shallow and the Deep, Long-range Ecology Movement: A Summary." *Inquiry*, 16, 95-100. Norton, Bryan G. (1991) *Towards Unity Among Environmentalists*. New York and Oxford: Oxford University Press. O'Neill, John (1993) *Ecology, Policy and Politics: Human Well-Being and the Natural World*. London: Routledge.

O'Neill, John, Alan Holland and Andrew Light (2008) *Environmental Values*. London and New York: Routledge.

Parfit, Derek (1984) *Reasons and Persons*. Oxford: Oxford University Press.

Passmore, John (1974) *Man's Responsibility for Nature*. London: Duckworth.

Plumwood (previously Routley), Val (1991) "Nature, Self and Gender: Feminism, Environmental Philosophy, and the Critique of Rationalism." *Hypatia*, 6.1, Spring, 3-27.

Rolston, Holmes, III (1975) "Is There an Ecological Ethic?" *Ethics*, 85, 93-109. Routley (later Sylvan), Richard (1973) "Is There a Need for a New, an Environmental Ethic?" *Proceedings of the XVth Congress of Philosophy*. Varna (Bulgaria), 205-10.

Routley, Richard and Val Routley (1975) *The Fight for the Forests*. 3rd edn., Canberra: Research School of Social Sciences, Australian National University.

Routley, Richard, and Val Routley (later Plumwood) (1978) "Nuclear Energy and Obligations to the Future." *Inquiry*, 21, 133-179.
Routley (later Plumwood), Val and Richard Routley (1980) "Human Chauvinism and Environmental Ethics," in Don Mannison, Michael McRobbie and Richard Routley (eds.), *Environmental Philosophy*. Canberra: Research School of Social Sciences, Australian National University, 96-189.
Scherer, Donald (1983) "Anthropocentrism, Atomism, and Environmental Ethics," in Donald Scherer and Thomas Attig (eds.), *Ethics and the Environment*, Englewood Cliffs, N.J.: Prentice-Hall, 73-81.
Shue, Henry (1995) "Equity in an International Agreement on Climate Change," in *Proceedings of IPCC Workshop, Nairobi, July 1994* (on "Equity and Social Considerations Related to Climate Change"). Nairobi: ICIPE Science Press, 385-92.
Singer, Peter (1993) *Practical Ethics*. 2nd edn., Cambridge: Cambridge University Press.
Singer, Peter (2002) *One World:the Ethics of Globalization.* New Haven, CT and London: Yale University Press.
Thompson, Janna (1990) "A Refutation of Environmental Ethics." *Environmental Ethics*, 12, pp. 147-160.
Tudge, Colin (1995) *The Day Before Yesterday*. London: Jonathan Cape.
Yearley, Steven (1996) *Sociology, Environmentalism, Globalization: Reinventing the Globe*. London, Thousand Oaks, CA and New Delhi: Sage Publications.

3

Moral Ambiguities in the Politics of Climate Change

Freya Mathews

Historically, the discourse of environmental ethics rested on the biocentric/anthropocentric distinction: an environmental ethic was one that extended moral significance beyond the sphere of merely human interests to the interests of life at large. However, in the context of climate change, this biocentric/anthropocentric distinction is becoming blurred, since efforts to mitigate climate change in the interests of human survival also generally serve the interests of the larger biosphere. In the context of climate change, in other words, environmentalism is undergoing globalization: its object is now the biosphere as a whole rather than particular ecosystems or species. However, there are different ways the biosphere as a whole may be conceptualized, and from these different conceptualizations, very different mitigation strategies, with very different consequences for other-than-human life, flow. It is important then, in the interests of environmental ethics in its original sense, as an ethic inclusive of other-than-human forms and systems of life, to distinguish these different conceptions of the biosphere and track their ethical implications.

Environmental Ethics and Anthropocentrism vs. Biocentrism

I happened to attend one of the very first screenings of *An Inconvenient Truth* in Australia. There was little advance notice of what we, the audience, could expect. When, in the stunned aftermath of the screening I walked out of the cinema, I remember thinking, "everything has changed." And it had. The year 2006 was the year the world woke up to climate change. This was not only because of Al Gore's film, of course; but the film certainly helped to precipitate the exact historical moment of awakening.

Although it was exhilarating to witness the moment of awakening, it was also, for me, touched with personal ambivalence. Ecological philosophy had been my life's work up to that point, and ecological philosophy had been, at its heart, a moral wake-up call. It had been essentially an argument for the moral significance of nature in its own right, over and above its utility for human purposes. It had, in other words, been an argument for some form of *biocentrism*, or *bio-inclusiveness*, in our moral thinking, or for a specifically *environmental* ethic, an ethic of nature.[1] Of course, there had been those who had argued that moral protection for the environment could be derived from the purely human-regarding ethical systems that had hitherto characterized the Western tradition, and that a new biocentric departure was neither required nor justified.[2] But insofar as ecological philosophy embraced the biocentric perspective and insisted on the intrinsic moral significance of nature, it did indeed inaugurate a new phase in the philosophical thinking of the West. The argument for this intrinsic moral significance of nature took many different forms, and solidified into many different positions, which were often at loggerheads with one another, but they were nevertheless basically all lined up against the blindness of Western civilization to the moral consideration of other-than-human forms and systems of life. The argument of ecological philosophy was, in other words, basically pitched against an anthropocentrism deeply entrenched in Western thinking since its inception.[3] The anti-anthropocentrism of the new environmental ethic, so counter to the basic Western mind set, articulated a sentiment that was core to environmentalism. Not all environmentalists shared it, but it was the main moral wellspring of the movement, and gave to the movement its distinctive moral passion.

The argument for the intrinsic moral significance of nature hinged on the question of what it is that entitles an entity to moral consideration. For many, though by no means all, ecological philosophers, an entity was entitled to moral consideration if it possessed an attribute of the kind that could broadly be described as mind-like. Although philosophers differed in their exact characterization of this attribute, most, though again by no means all, shared a tacit assumption that the concept of nature had in the West been constructed in opposition to the concept of humanity, and that this dualistic construction had reserved mind for us and had drained all mental-type qualities out of nature, leaving nature with only the matter-qualities accorded it by a strictly materialist science. By mental-type qualities, I mean here not only consciousness as it is found in the human instance, but larger possibilities of intelligence, subjectivity, sentience, agency, intentionality, telos, or conativity, such as might belong to a wider range of organisms, and perhaps even to all of life.[4] In any case, it is manifestly such mentalistic qualities, in this larger sense, that imbue an entity with a degree of self-meaning, a degree of mattering to itself. And an entity which matters to itself is categorically different, from a moral point of view, from one which does not: destroying an entity that matters to itself is, at least prima facie, a different proposition, morally speaking, from destroying one which does not matter to itself. There is, in other words, a fundamental relationship between morality and mattering, and while it may indeed be that entities can only have moral significance if they matter to some-one, that someone can, in the case of entities with mental attributes, be themselves. In order to show that nonhuman as well as human entities are morally considerable then, ecological philosophers often argued, contra mind-matter dualism, that nonhuman existence, like human existence, is permeated with mind-like attributes.

Restoring mental attributes to nature in this way and accordingly extending moral consideration beyond the circle of human interests represented, as I have remarked, a radical challenge to the metaphysical and moral premises of Western civilization, with its foundational commitment to a materialist science and an instrumentalist economism. It was for this reason that the discourse of ecological philosophy, with its advocacy of a specifically environmental ethic, or ethic of nature, had a dissident status, and to the extent that this discourse informed environmentalism, lent the movement a counter-cultural flavor.

Precisely because the roots of anthropocentrism lay so deep in Western thought, few of us ecological philosophers or environmentalists were, I think when I look back, really surprised that our call for a new environmental ethic was largely ignored by mainstream society. We understood that such an environmental ethic was morally revolutionary in its implications for the West, and was hence at best a long term goal. So some of us were a tad unprepared when suddenly, in 2006, the great wake-up seemed to be upon us! After decades of denial, society was suddenly facing up to the frightening urgency and gravity of the climate crisis. And the climate crisis was so huge in its environmental implications it seemed to swallow up other environmental issues—environmentalism itself seemed to morph, in the public imagination, into response to climate change.

Welcome as this sudden green awakening was, I personally was left, as I have remarked, wondering what further role, if any, there would be for ecological philosophy. Now that it had been widely acknowledged that we were in the midst of an unprecedented environmental crisis, the ball was, it seemed, in the court of economists, scientists, and designers: their task was to work out, in practice, how to re-design the production regimes of modern societies to bring them in line with sustainability requirements. Moral musings seemed, in the new Titanic context, superfluous. However, I held my peace, and waited to see how the dust would settle. I did notice that, at my own university, nobody beat a path to my door and said, "Oh, you ecological philosophers were right after all!" Nobody apologized for the marginalization that ecological philosophy had always endured within the academy. I noticed too that despite the fact that debate on climate change was dominating the airwaves, the new "experts" were not the old environmentalists and greenies, but the same economists, managers, industry leaders, and politicians who had but a few years earlier scoffed at the very idea of an environmental crisis. They seemed to be taking the newly acknowledged crisis more or less in stride, a new variable to be factored into the old equations. I fully acknowledged that practical solutions were indeed what we needed now, and that practical solutions would emanate from economists, scientists, business leaders, lawyers and the like. But nevertheless, something seemed to be missing in the new debate.

Having accepted that environmental problems were extremely urgent and serious after all, mainstream commentators and leaders seemed to imply, in their behavior, that such problems could certainly not be entrusted to those pipe-dreaming counter-culturalists, the greens. The time had come, it appeared, for proper hard-headed solutions to be brought forward. This was code for more of the same thinking that had produced the crisis. In other words, the value system seemed not to have changed. It was as anthropocentric as ever. The whole "awakening" to the environmental crisis in the shape of climate change had in fact pretty much by-passed the moral question. Yet the climate change debate was clothed in the moral rhetoric of environmentalism, and for this reason environmentalists appeared to accept its legitimacy.

Nature of Climate Change and Moral Distinction between Anthropocentrism and Biocentrism

But why did environmentalists, or those of them committed to an environmental ethic, not see through this? Why did they seemingly, by and large, accept the terms of the climate change debate, and welcome it? One reason, I think, lies in the way in which the nature of climate change itself *masks* the moral distinction—between anthropocentrism and biocentrism—on which an environmental ethic rests. Climate change introduces moral *ambiguity* into environmental debate. Or that, at any rate, is the proposition I wish to explore here.

I shall consider two main ways in which climate change introduces such ambiguity. First, there is the matter of *scale*. The scale of the climate problem is so great as to simply, potentially, overwhelm environmental ethics: when the habitability of the planet per se is at stake, efforts to maintain its habitability for humans will *ipso facto* help to keep it habitable for all life systems. In past environmental struggles, human interests and the interests of the other-than-human world have often been locked in conflict; the moral issue in these cases has been whether we are justified in appropriating or destroying nonhuman beings or systems in pursuit of our own self-interest. In the current climate change crisis however, human interests and the interests of other-than-human entities and systems seem to be thrown together into the same life boat: in order to protect itself, humanity has to secure the physical conditions for life per se, and in doing so, it will incidentally be helping to save the rest of nature. The old biocentric/anthropocentric distinction thus appears in this context to no longer be useful.

Although there is some truth to this claim—that in the climate change context our efforts on our own behalf will also, incidentally, benefit the rest of nature—our choice of actual mitigation strategies will be profoundly influenced by our moral presuppositions. If these presuppositions are anthropocentric, our goal is likely to be to perpetuate industrial civilization in basically its present form by relying on a purely instrumental scientism for our solutions. This will involve devising large-scale mechanistic or engineering remedies that are conceived with little regard for their consequences for the rest of life. Opting for nuclear power as an alternative to fossil fuels is an example of this approach. So too are the various large-scale engineering strategies that have been proposed to mitigate global warming. These include the use of giant reflectors in space to regulate atmospheric temperature or the pumping of sulfur into the atmosphere to achieve solar dimming and thereby, again, reduce global temperature. Other such suggestions include firing particles into the stratosphere to replicate the effect of "volcanic winters" and pouring chemicals into the oceans to encourage the growth of algae that will then gobble up CO_2 from the atmosphere. The rush to cut into the last of the earth's forests and shrublands for bio-fuel production follows a similar anthropocentric trajectory, advocating use of the last remaining biological resources of the planet to prop up our present industrial and transport regimes.

In all these cases, a science which treats nature as pure mechanism is brought to bear on the problem of thermal instability, and narrowly-conceived solutions which serve industrial civilization are proposed with little or no regard for their consequences for planetary life-systems. In other words, without a conscious commitment to biocentrism, Western approaches to climate change are likely to substitute technical systems for natural ones and to select mechanistic strategies that impact negatively on natural systems. Even if these strategies succeeded in securing the physical conditions required for climate stability—and it is doubtful that they would do so—their own consequences may turn out to be as harmful to other-than-human life systems as climate change would have been.

While we might then concede that *scale* alone does not collapse the moral distinction between an anthropocentric and a biocentric course of action in the context of climate change, there is a *second* factor arising out of this context that also generates moral ambiguity. This is the factor of *globalism*. Since climate change is inherently and irreducibly a global phenomenon, it forces us to re-cast environmentalism in global terms. Insofar as the moral object of environmental ethics has traditionally been "nature," climate change forces us to think about nature not in terms of particular ecosystems or organisms or even species—this forest, that wetland, this endangered butterfly, that significant habitat—but rather in terms of the biosphere as a whole. Nature, in other words, is understood holistically rather than as a mere aggregate of living things: it is the planetary system as a whole that generates climatic phenomena. But how exactly is this whole to be conceptualized?

I would like to propose two ways—both holistic, but differently so—in which "nature" under its global aspect might be conceived. The first such way in which nature might be conceived is as a self-realizing or autopoietic system, defined not in terms of the elements that contingently constitute it, but in terms of its ends as an entity in its own right, which is to say, in terms of its status as an end-for-itself, and its disposition to navigate circumstances in such a way as to preserve its own identity as a living system through time and change.

The second way in which nature under its global aspect might be (holistically) conceived is as a self-realizing or autopoietic system, yes, but one which is defined not merely in terms of its ends—the end of self-preservation—but also in terms of its specific pattern of organization, its pattern of self-structuration. The moral significance of nature under its global aspect, from this point of view, lies as much in this pattern of self-structuration as in its status as an end-for-itself. In protecting it, we would not only preserve its physical continuity through time and change, but its particular organizational integrity as well.

For a fuller appreciation of the distinction between these two interpretations of global nature, let us examine each of them in turn.

If, in speaking of the biosphere as a whole, we intend to indicate a Gaian-type self-regulating or autopoietic system which seeks actively to preserve and promote the physical conditions necessary for its own self-perpetuation, then it might seem natural to assume that in protecting that system we will also be protecting the particular ecosystems—the forests, wetlands, species, habitats, and individual organisms—that make it up. We might assume, in other words, that our efforts on its behalf will have outcomes consistent with biocentrism. But is this necessarily the case? Consider an analogy. An organism, let us call it O, might evolve a capacity to replace the organs in its body with crafted structures that reproduce the functional effects of the original organs. O might undergo a quite extensive internal conversion of biological organs to replacement devices that nevertheless ensure O's perpetuation just as effectively as the original organs did. O's own substitution of devices for organs might well then be consistent with its original autopoietic ends—this substitution would not necessarily detract from O's status as an end-for-itself and as a self-realizing system, though the "self" that it preserves and realizes is not the original community of organs but a hybrid of organs together with devices of its own making. We ourselves are, after all, such an organism, capable of substituting technological devices for some, maybe one day many, of our organs, yet remaining dedicated to our own self-realization and in this sense retaining our commitment to our original autopoietic ends.

If nature under its global aspect is defined purely in terms of its autopoietic ends, there seems no reason, morally speaking, why we should not substitute technology for ecology as the internal mechanism for the self-preservation of the biosphere. "Biosphere" here signifies a system that reproduces the conditions for life, where "life" is understood in generic rather than specific terms, as a physiological process rather than as any particular pattern of species composition. That we ourselves ended up more or less exclusively occupying the resultant space for life would no more count against the autopoietic status of the planetary system, from this point of view, than multiple organ replacements would count against the autopoietic identity and continuity of the human patient.

In other words, if nature under its global aspect is defined purely in autopoietic terms, then "nature" might well be "saved" by our substituting technological systems for many or most ecological ones. There is no telling, a priori, that this is not possible—that we might not be able to engineer the planet so that it remains hospitable to "life"—in the shape of ourselves—even though most natural ecosystems have ceased to exist. To put this in Gaian terms, as products of Gaia's self-activity, our activities, including their technological expressions, might be seen as just as integral to Gaia's autopoietic agency as are the activities of biotic systems. If Gaia's moral significance lies in its status as an autopoietic system, then can't we humans see ourselves as agents of Gaia's moral purpose? But saving "nature" by substituting technology for ecology is hardly the kind of outcome intended by proponents of a bioinclusive environmental ethic.

Now let us consider the position that what is morally significant about the biosphere under its global aspect is not merely that it is a living system in the autopoietic sense—though we may be agreed that it is indeed a living system in this sense—but that it is a living system in a further sense, namely inasmuch as its internal organization conforms to certain definitive principles. In other words, what makes the biosphere as a whole morally considerable, according to this position, is not just that it is an end-for-itself and is self-active in pursuing its own continued existence, but that its way of pursuing its existence—the strategies and principles of organization it follows in doing so—represent the embodiment of what might be called a special *genius* that is the peculiar province and merit of living things. This genius is characterized by the economy of means that living things employ to achieve their ends—their ability to convert wastes from other organisms into resources for themselves and to make of their own costs opportunities for others, and vice versa. In other words, living things realize themselves by availing themselves of the resources provided gratis by others, where this has the consequence that their possibilities of self-realization are tightly shaped by one another. Such an economy of means is, as the turning of costs into opportunities for others demonstrates, logically tied up with a functional interdependence so intricate as to make such systems effectively mutually inextricable.

This relational inter-functionality, integral to the peculiar genius of living systems, gives such genius a distinctively moral texture, a texture of accommodation of self to others, that confers on living systems a moral status additional to the moral status that accrues to them simply on account of being ends-for-themselves. The genius of life is then a genius of adaptation of self to other, which is at once a genius of intelligence, insofar as it conserves effort and thereby increases fitness, and a moral genius, indeed the very prototype of morality. Since the biosphere as a whole is a system internally organized in accordance with this principle of accommodation of self to other, and moreover constitutes the condition for the like organization of its component systems, it is the pre-eminent locus of this peculiar genius and of the merit associated with it, and for this reason may deserve our moral consideration.

If the biosphere, viewed as a locus of such genius, is the object of our moral concern, then our own interventions on its behalf have to conform to its distinctive pattern of organization. It would not suffice, in this instance, merely to secure the physical conditions necessary for life, in a generic sense, on the planet. We would have to secure those conditions by the same kind of means that nature does, for otherwise our interventions would destroy the very thing—the pattern of organization—we seek to preserve. This way of interpreting nature indicates the path of biomimicry, biomimicry being a philosophy which takes nature as the model for design and organization in human praxis. As biomimicry advocate, Janine Benyus, explains, biomimicry "is a new science that studies nature's models and then imitates or takes inspiration from these designs and processes to solve human problems" (Benyus 1997, xi). Biomimicry points more generally to the idea of a "circular economy," a closed-loop system that converts all "waste" incurred in (agricultural or industrial) production into resources for further production. ("Circular economy" is the term adopted in China for industrial ecology, or the organization of industrial production in accordance with the principles that structure ecosystems in nature. It is defined in opposition to a "linear economy," which rests on the assumption that resources can be secured in endless supply at one end, and allowed to build up in the environment as waste at the other end.[5])

However, although biomimicry is undoubtedly a key concept in addressing climate change, and the circular economy is a key development in sustainability theory, neither of these will necessarily, of themselves, lead to bio-inclusive outcomes. For suppose it turns out that we can indeed re-design industries and transport-systems and built-environments along strictly biomimetic lines. We might discover how to construct "living buildings," for instance, which, in the process of serving the needs of their occupants, capture solar energy, precipitate and circulate water, extract oxygen from carbon dioxide, and regulate their own temperature purely via physical design.[6] We might design industrial plants that purify and recycle water as they consume it and filter and regulate atmospheric gases, while at the same time, with their waste streams, create resources for further manufacturing and for food production.[7] We might eventually become so adept at designing cities and industries and transport systems that functionally equate with ecological systems that the need for ecological systems as mechanisms of planetary self-regulation will be obviated. Solar cities that photosynthesize may take the place of forests, and industrial "plants" that purify and reticulate water may take the place of wetlands. Manufacturing processes that include food in their outputs and confine production inside closed resource loops may replace traditional agriculture and bypass the need for resource extraction and waste disposal. The physical conditions for life may, in other words, ultimately be maintained by biomimetic technical systems that render superfluous the biological systems they imitate, with the result that the "planetary life" which these conditions safeguard becomes vested exclusively in us.

The Need to Protect the Biosphere as a Whole

In relation to both interpretations of global nature, it may be argued that seeking to protect nature under its global aspect subtly tends to prioritize the interests of the system as a whole over the interests of its component parts. Whether this system is identified in terms of its autopoietic ends, as in the first, or its distinctive pattern of organization, as in the second, we could protect it—its ends or its pattern of organization—while sacrificing its actual living components, the forests, wetlands, species, habitats, and individual organisms that are the objects of a biocentric ethic. In other words, the apparently biocentric goal of "protecting nature," in the global context of climate change, could quite conceivably lead to outcomes which were anthropocentric.

So the issue of climate change is fraught with moral ambiguity for environmentalists. I would suggest that the way forward is to maintain our traditional environmental commitment to local ecosystems as moral ends-in-themselves, and not allow this commitment to be subsumed under the new global perspective emanating from climate change. In other words, concern for the biosphere as a whole needs to be balanced with concern for its component parts if we wish to retain a biocentric orientation. As long as earth is valued exclusively as a global system, its parts may be sacrificed for the sake of the whole. In making this point I am re-playing earlier debates in environmental ethics regarding the rights of ecosystems versus those of individual organisms.[8] However, the moral ambiguities that vex questions of part-whole relations do arise again in the context of climate change, and it is important to be alert to them if we wish to avoid conflating biocentric and anthropocentric objectives.

That said, it is at the same time self-evident that environmentalists committed to bio-inclusive outcomes also need to address the problem of climate change, since stable climatic conditions are as necessary for nonhuman life as they are for human life. So how can environmentalists of a biocentric persuasion maintain their moral commitment to particular ecosystems while still taking into account the global requirements of climate stability?

Bio-Synergy Will Protect the Biosphere and Address the Climate Change Problem

I would suggest that we rely on these ecosystems themselves as our main means of climate change mitigation and as our main avenue to sustainability in production. I am suggesting, in other words, that we maintain our traditional commitment to natural ecosystems while also discovering how to make these systems work for us, as never before. This means entering into a partnership with ecosystems, being prepared not merely to conserve them in their present form, but to act as agents of their evolution, adapting and changing them, but in directions that are consistent with their own biotic integrity and conative trajectory. I call this kind of partnership *synergy*.

Since conativity and synergy are salient categories here, let me take a moment to explain them. By the term *conativity*, I mean the innate impulse of living things to maintain and increase their own existence, each in their own particular mode or style.[9] By *synergy*, I understand a form of relationship between two or more conative parties who engage with each other in such a way that something new and larger than either of them, but true to the conative tendency of each, is born. Synergy is a modality which brings novelty and change into the world but in a way that is consistent with the conative grain of things at any given moment in the self-unfolding of nature.[10]

To engage synergistically with natural ecosystems then, in the interests of both climate change mitigation and sustainable production, would mean harnessing their energies while yet ensuring that our interactions with them did not revert to the one-way instrumentalism of anthropocentric regimes. For this to be feasible, we would need to possess vastly expanded insight into the nature of these systems, and to be prepared to adapt our productive requirements, and hence our own self-expressiveness, to the conative contours of biological systems. Let us look at the implications of these two conditions.

To understand biological systems to the degree necessary for making them our principal means of both climate change mitigation and sustainable production would mean a significant expansion of traditional biological and ecological sciences, and the addition of new kinds and methods of "science" that would offer insight into the conative tendencies of organisms and living systems.

In order to explain, let me offer an example of the kind of expansion of traditional biological science needed to appreciate the role of ecosystems in climate change. It is already well known how forests, for example, function as carbon sinks, absorbing carbon from the atmosphere and storing it in forest soils (Garnaut 2008, 35-36; Mackey, Keith, Berry and Lindenmayer 2008). But new work is pointing to how little we have in fact to date understood the biospherical functionality of forests, and hence how ill-equipped we have been to take informed biological action in response to the climate crisis. The work in question emanates from a small group of scientists in Australia, the Sustainability Science Team of the Nature and Society Forum in Canberra, led by Walter Jehne. They offer a somewhat heretical analysis of the causes of climate change (Jehne 2007). Not being a scientist myself, I cannot vouch for their claims. I am introducing these claims here rather to illustrate, in a small way, just how under-developed current science may be in its account of the contribution of biology to climate stability.

In his analysis of climate change, Jehne points out that though the ice core data reveal an association between increased levels of atmospheric carbon and elevated temperatures over the past 600,000 years, the increase in carbon often precedes the increase of temperature by thousands of years. This suggests that carbon is not in fact causing temperature increase, but rather that carbon increase and temperature increase are effects of a common cause. In any case, Jehne explains, carbon is a relatively small player in the thermal regulation of the atmosphere: the main player is water. It is the amount of H_2O in the atmosphere that accounts for 95 percent of thermal flux, while other gases account for only 5 percent.[11] Jehne then hypothesizes that the past variations in atmospheric temperature and CO_2 evident in the ice core data are common effects of changes in the hydrological cycle of the planet. Optimal H_2O in the atmosphere produces optimal thermal conditions for life by ensuring cloud cover with the right albedo (capacity to reflect solar radiation). Optimal H_2O in the atmosphere is maintained by vegetation—by the native forests and shrublands and grasslands that emit vast quantities (over a billion tons annually) of microbial life into the atmosphere. The microbes act as microscopic aerosols that afford condensation surfaces and nucleate water vapor into micro-droplets that play a major role in cloud formation. When land is cleared, and native vegetation with its cloud-forming function destroyed, cloud cover is diminished, and the amount of solar radiation reaching the surface of the earth increases.[12] Atmospheric CO_2, of course, also increases as a result of the drop in bio-sequestration of carbon that accompanies land clearing.

Jehne hypothesizes that the observed current increase in both atmospheric temperature and atmospheric CO_2 is the result of the vast scale of land clearing that has occurred since the beginning of industrialization two hundred and fifty years ago, rather than of the industrial carbon emissions of the last thirty years. (He reminds us that the thermal effects of these latter emissions are actually yet to be seen, as carbon from emissions is absorbed by the oceans for about fifty years before it finally finds its way back into the atmosphere.) Of course, there is nothing heretical about including land clearing as a source of carbon emissions in climate change computations. The difference in the approach of the Sustainability Science Team is that land clearing is seen as significant not so much for the carbon emissions it produces as it is for its effects on the hydrological cycle. Their argument is, as I have indicated, that it is primarily alterations in the hydrological cycle rather than in the carbon cycle that are currently driving climate change. They assert that "a 1% increase in mean solar reflectance through increased cloud albedos may have an equivalent effect in cooling the earth's surface climate to that of reducing current CO_2 levels back to preindustrial levels" (Jehne 2007, 5).

If the destruction of forests and other native vegetation systems on a planetary scale is the major cause of global warming and climate destabilization, then the "solution" may simply be to allow those forests and other vegetation systems to recover. All that is required of us in the present situation of crisis then is to allow forests and other ecosystems to follow their own conative trajectory and regenerate. Reduction of greenhouse gas emissions of course remains important, but the principal task is regeneration. As the hydrological cycle is readjusted by such regeneration and the climate problem thereby ameliorated, carbon will also be sequestered, soil loss will be reversed, soil structure improved, salination and desertification redressed, rivers revived, water supplies replenished and purified, and biodiversity boosted as habitat is restored across the planet.

This, then, is a biological solution to climate change, and it requires very little of us, not even synergy. It requires only that we give full rein to the conative tendencies of planetary vegetation systems, just as a biocentric ethic dictates that we should. Of course, for us to allow forests and other vegetation systems to get on with the job in this way does entail major readjustments of our own economic praxis, for it involves giving back to nature, so to speak, land and resources that have been used by us for extractive, manufacturing, agricultural, and other intensive, bio-destructive purposes. We will accordingly have to meet our productive needs in other ways, ways that are consistent with nature meeting its needs. I shall take up this point shortly.

For the moment, let me return to the second way in which science has to be expanded if we are to balance the moral claims of the biosphere as a whole with the moral claims of its component parts, as required by a biocentric ethic. I suggested that such a balance could be achieved if we were able to rely on these component biological systems as our means of both climate change mitigation and sustainable production. We have already seen how climate change mitigation could be achieved simply by allowing biological systems to recover. Affording the conditions for such recovery, once we have understood through an expanded science why this is necessary, requires little in the way of synergy and a minimum of insight into the details of the conativity of these systems. To effect wholesale mobilization of biological systems in the service of sustainable production, however, is a different matter. It would require complex synergies with these systems and accordingly detailed knowledge of their conative tendencies. Only with such detailed knowledge could we be sure that our interactions with them were genuinely synergistic, which is to say, consistent with their own inherent tendencies, rather than purely instrumental, and hence anthropocentric.

But how can we attain such knowledge of the inherent tendencies of living entities? We can appreciate that the basic conative tendency of each living thing is to maintain its integrity and repair or regenerate itself when damaged. But how can we discover the distinctive *style* of self-realization that constitutes the conative essence of any particular entity, and thereby circumscribe the transformations that are consistent with its integrity? To discover this distinctive style of self-realization is of course critical to attaining synergy, since synergy is defined in terms of it. But traditional science offers little or no insight into styles of self-realization, since such phenomena are not reducible to the analytical terms of traditional science. This is because a living thing's style of self-realization emanates from the inner unity of meaning that a thing has for itself rather than from the underlying laws of physics. This inner unity of meaning, which constitutes the living thing's own sense of itself, is expressed in the patterns whereby it seeks to actualize itself, patterns that are selected out from the unlimited possibilities arising from mere physical causation by the inner meaning these patterns hold for the thing in question (Mathews 2008). To discover such underlying patterns of meaning does indeed require empirical methods, but empirical methods that lie well outside the repertoire of traditional science.

One method that might serve this purpose is the method of Goethean science. Goethe, the eighteenth century poet and naturalist, outlined a four-step procedure (exact sense perception, exact sensorial imagination, seeing-in-beholding, and being one with the object) that started with contemplative observation of an entity, but opened out into a form of communicative engagement with it that involved the exercise of carefully disciplined faculties of intuition and imagination as well as perception in order to discover the distinctive "gesture" of the entity that was expressed, but never entirely articulated, in the appearances it presented to observers (Brook 2009, Bortoft 1996). It would take me too far afield to explain this procedure in more detail here. Suffice it to say that if synergy with living systems is to be the basic modality of a new civilization indicated by a biocentric ethic in the era of climate change, new forms of cognition or epistemic protocols in addition to those countenanced by traditional science will need to be developed.

Now, finally, let us consider the proposition that biological systems should provide our principal means of production as well as our principal means of climate change mitigation. This proposition was put forward in response to the biocentric requirement that we achieve climate stability without compromising the integrity of the component entities or systems that make up the biosphere. To satisfy the proposition, I suggested, would involve harnessing natural ecosystems for the purposes of production without reducing them, instrumentally, to mere means of ours. This could be achieved by synergy, by allowing natural systems to serve our ends, but only to the extent that their doing so was compatible with the pursuit of their own ends. Where the ends of ours contradicted the conative tendencies of natural systems, those systems could not be conscripted by us. Instead our ends would have to be adapted to the conative contours of the systems.

Let us describe an economy based on this kind of relationship with biological systems as a *bio-synergistic* one.[13] Bio-synergy is clearly distinct from biomimesis in that in a biomimetic economy technology is designed in accordance with biological principles of organization, whereas a bio-synergistic economy operates in actual partnership with existing biological systems. Bio-synergy will undoubtedly overlap in practice with biomimesis, but the two are conceptually distinct in important ways. The main difference in their perspectives is that for bio-synergy, sustainability is a matter of meeting human needs while assuring the viability of existing life systems, whereas biomimesis aims to reproduce, in the processes of production, the conditions for life in a generic sense, without pre-specifying the ultimate membership of the life community.

A bio-synergistic economy brings aspects of a forager economy into the twenty-first century in that it relies on provisions from the bio-energy systems that are already available in the biosphere rather than replacing these systems with agricultural and manufacturing systems of its own. Any such bio-synergistic economy, understood here merely as an ideal type rather than as a presently feasible alternative, will include certain cardinal aspects.

Such an economy will be, first and foremost, a solar economy, since solar energy animates the entire fabric of natural biological systems and can be gathered without cost to those systems.

Bio-synergy will, secondly, involve adjusting human demand to ecological carrying capacity, where ecological carrying capacity is understood to mean the capacity of ecological systems to support human populations without compromising other-than-human constituencies. Biosynergy in this respect is patently incompatible with current levels of human population and therefore prescribes human population decline.

Thirdly, bio-synergy indicates that instead of practicing traditional agriculture, we should allow indigenous ecosystems to serve as our primary producers. "Bush foods" (or, in the Australian context, "bush tucker") will in this sense constitute staples in a bio-synergistic economy, though it is imperative to qualify this statement with the condition that bush foods will only be harvested to the degree required for the regulation of ecosystems. In other words, the role of human consumers in the ecosystem will replicate that of other predators, routinely reducing populations of consumed plant and animal species to ecologically optimal levels. (It is paramount to state this qualifier since the commercial harvesting of "bush meat" in economies in which nature is already under attack is often the last nail in the ecological coffin.)[14] In countries like Australia where feral species—plant and animal alike—pose major threats to indigenous biological systems, ferals will be the first targets of the new biosynergistic regimes of organized foraging. Bypassing such species as objects of consumption is one of the most striking anomalies of present bio-antagonistic economies. Australia, for instance, is host to vast populations of invasive feral animals, such as rabbits, goats, pigs, and camels, yet these animals almost never appear on the national table. Further ecological damage is incurred, on an even vaster scale, to deliver traditional farmed animals—sheep, chickens, and cattle, for instance—to the table instead. Readiness on the part of consumers to switch from traditional meats to feral meats and ultimately to indigenous meats, in reduced quantities, provides an example of the kind of adaptability required of consumers in a biosynergistic economy, in which two-way accommodation of ends is expected. It must be remembered, in other words, that bio-synergy is a two-way street—it allows us to act on nature, but it also permits nature to act on us, trimming our ends, and with them our self-expressiveness, to the conative contours of ecosystems.

Insofar as we rely for provisions on the bio-energy systems that are already available in the biosphere rather than replacing those systems with agricultural and manufacturing systems of our own, we are exemplifying the forager aspects of bio-synergistic economies. However, biosynergy is not exclusively a forager modality. It allows us not only to gather produce from pre-existing biological systems, but also to proactively modify those systems, at least to the extent that such modifications represent a further self-unfolding of those systems rather than their thwarting. So, for instance, we might vary the physical conditions that define the parameters of particular ecosystems, thereby changing those systems, but in a direction we judge to be consistent with their conative tendency. We might, for example, increase the number of water holes in an arid region, thereby increasing the density of wildlife there. Such an action would need to be offset by other actions, since increasing the density of wildlife in a fragile arid ecosystem might well incur degradation of the vegetation. But a judicious selection of actions might result in the "evolution" of the original ecosystem into one with higher biomass and higher biodiversity than the original system. If we judge the conative tendency of living things to include the impulse not only towards self-maintenance but also towards self-increase, it is arguable that such a change in an ecosystem—in the direction of greater vitality and diversity—would qualify as consistent with the system's own conativity. In this sense, it may be possible for us to intervene in biological systems to increase their "productivity" without violating their integrity. Clearly such complex and delicate interventions, which at the same time serve our own interests and the interests of the systems themselves, would only be possible to the extent that we fully understand the physical and conative capacities of systems.

Primary production in a bio-synergistic economy then might be figured as a sophisticated and proactive custody of indigenous ecological systems. In areas already laid waste by traditional farming, such systems could be re-introduced. More intensive forms of boutique food production, featuring the full range of available food species, could also be integrated into industrial and urban design, thereby taking pressure off natural systems. But industrial production in a bio-synergistic mode is more difficult to envisage. Certainly manufacture can take advantage, biomimetically, of natural materials and energy sources; transport systems can piggyback on natural carriers, such as rivers, oceans, and wind flows. Architecture can exploit topography and siting to minimize the impact of buildings on landscapes. But the kinds of strategies available to manufacture are primarily biomimetic rather than biosynergistic: it is hard to see how natural biological systems could, even in synergy with us, produce cars and kettles, let alone airplanes and computers. For the time being then, biomimesis, together with the progressive tailoring of our desires to the capacities of natural systems, might have to suffice: we might have to be content with a manufacturing system that functions biomimetically, without further waste or extraction, on a material resource base already carved out by industry, rather than looking to the agency of actual biological systems to take the place of industry. In the future, however, we might indeed achieve the purposes currently served by articles such as cars and kettles by harnessing the agency of natural systems more immediately and processually, without the need for clunky permanent articles of this kind. From the vantage point of such a biologically sophisticated future, we might look back on our present era of manufacturing as a kind of Dark Ages, an age of obtuse unnecessary clutter, blocking, short-circuiting, and destroying the elegant pathways of agency and efficacy already available in the shape of natural biological processes and systems.

Conclusion

In conclusion, I have argued that the way forward for bio-inclusive environmentalism in the era of climate change is not merely *biophysical*, in the Gaian-type sense, involving the engineering of the physical conditions requisite for life on the planet; nor merely *biomimetic*, in the sense of replicating, in our technologies, the patterns of organization characteristic of life; but *bio-synergistic*, in the sense of entering into active partnership with actual ecosystems to ensure both the regulation of the climate system and the sustainable provision of our own needs.

I am under no illusions that, as a political goal, bio-synergy is achievable at the present time; but I think it is our responsibility, as ecological philosophers and thinkers, now as before, or now more than ever before, to keep this goal visible in the public debate.

Notes

1. Early architects of this specifically environmental ethic, or ethic of nature in its own right, included Routley, 1973: 205-10; Routley and Routley, 1980, 96-189; Arne Naess, 1973, 95100; Rolston III 1975, 93-109; J. Baird Callicott, 1979, 71-81.
2. This was argued most famously, in the early days, by Passmore 1974.
3. Feminists, particularly ecofeminists, provided particularly telling analyses of the way Western thought was organized around a category of nature dualistically opposed, and inferior, to all that went to make up our concept of humanness. Out of such a dualistic system of thought, anthropocentrism emerged as our natural and legitimate standpoint vis-à-vis nature. Early analyses of this kind included Griffin 1978; Daly 1978; Ruether 1975. Later, less essentializing, more critical treatments of this theme included Lloyd 1984; King 1990, 106-121; Plumwood1993.
4. For classic discussions of criteria of moral considerability, see Rodman 1983, 82-92; Goodpaster 1978, 308-325; Taylor 1981, 197-218.
5. John Mathews, Professor of Management, Macquarie University, personal correspondence. China adopted the circular economy as its official development model in August 2008, and is the first country to have done so. (The *Circular Economy Promotion Law* was adopted by the fourth session of the 11th People's Congress in Beijing in August 2008, as reported by the *ChinaDaily* in September.)
6. See, for example, Berkebile and McLennan, who discuss buildings designed to "generate a significant portion of [their] power without pollution, clean all [their] own wastes on site, and respond actively to temperature changes to maintain a comfortable indoor environment" (Berkebile and McLennan). Another example of thermal self-regulation in architecture is the Eastgate Building in Harare, Zimbabwe, an award-winning office complex designed according to the same thermodynamic principles as a termite mound. See *www.asknature.org*, the Web site for the Biomimicry Institute, founded by Janine Benyus. [7] See, for example, the brewery near Tsumeb in Namibia that has been designed for zero emissions. The brewery produces mushrooms, earthworms, chickens, spirulina algae, and fish in addition to beer. It generates fuel for its own operations and recycles all the water it uses (Mshigeni 2001; Saunders 2000).

8. This debate was famously played out between advocates of animal ethics and advocates of ecological ethics, the animal advocates emphasizing the rights of individual organisms and the ecological advocates emphasizing the over-riding ethical claims of ecosystems. See, for instance, Callicott 1980: 311-328; Callicott 1989: 49-59; Mark Sagoff 1984: 297-307.
9. I derive the term from Spinoza, who defines *conatus* as that endeavour whereby things strive to maintain and increase their own existence. See Prop VI, Part III of the *Ethics*, any edition.
10. I am using the term *synergy* here in a way that contrasts, to some extent, with its definition by comparable theorists. Peter Corning, for instance, uses the term to refer to "the combined, or cooperative, effects produced by the relationships among various forces, particles, elements, parts or individuals in a given context." (Corning 2003, 2) For Corning, the bricks in a wall and the atoms in a molecule are engaged in synergy, whereas in my sense, synergy implies a certain transformation in the participating parties, a transformation that represents a further elicitation of potentials inherent in their respective conativities. Synergy in this sense is not merely a cooperative way for agents to achieve their respective pre-elected ends but rather an interaction that transforms those ends, or redirects the agency of the parties, though in a way that enhances their self-realization. For a discussion of synergy in this sense, see Mathews 2006, 85-114; for a more metaphysical treatment, see Mathews 2003.
11. John Schooneveldt, a member of the Sustainability Science Team; private communication.
12. Although I cannot in any way vouch for the validity of Jehne's approach to the causes of climate change, it is worth noting that the Garnaut Report acknowledges the lack of understanding of the role of water vapor in thermal regulation: "[T]he lack of understanding of the way water vapour will respond to climate change, specifically its role in cloud formation, is a key factor in the uncertainty surrounding the response of the climate to increased temperatures" (Garnaut 2008, 34).
13. I am using the term *bio-synergy* here as a special instance of synergy, namely the kind of instance that occurs when humans engage synergistically with biological systems with specifically instrumental—though still synergistic—intent. A quick google of the term *biosynergy* threw up at least one author who was using the term in a cognate, but not quite identical, way. Anthony L. Rose of the Biosynergy Institute defines bio-synergy as "the fundamental and overarching process that sustains life on Earth. Its seed and its spark have been born and imbedded in every living cell, organism, and ecosystem since the beginning of time. Biosynergy is the inner force that compels each and every individual being to collaborate with others for the greater good" (Rose 2007).
14. For an account of the bush meat crisis—the large-scale commercial butchering of wildlife, including gorillas, chimpanzees, bonobos, and elephants, for domestic and export markets—see the Canadian Ape Alliance Web site as well as Anthony Rose's Web site, <bushmeat.net>.

References

Benyus, J. (1997) *Biomimicry: Innovation Inspired by Nature*. New York: Harper Collins.

Berkebile, B. and J. McLennan "The Living Buildings: Biomimicry in Architecture, Integrating Technology with Nature." Retrieved on January 2009 from <http://elements.bnim.com/resources/livingbuildingright.html>.

Bortoft, H. (1996) *The Wholeness of Nature: Goethe's Way of Science*. Edinburgh: Floris Books.

Brook, I. H. (2009) "Dualism, Monism and the Wonder of Materiality as Revealed through Goethean Observation." *PAN Philosphy Activism Nature*, 6, forthcoming.
Callicott, J. B. (1980) "Animal Liberation: a Triangular Affair." *Environmental Ethics*, 2, pp. 311-328.
Callicott, J. B. (1989) "Animal Liberation and Environmental Ethics: Back Together Again," in J. Baird Callicoott, *In Defense of the Land Ethic*. Albany NY: SUNY Press, 49-59.
Callicott,J. B. (1979) "Elements of an Environmental Ethic: Moral Considerability and the Biotic Community." *Environmental Ethics*, 1, 1, 71-81.
Corning, P. (2003) *Nature's Magic: Synergy in Evolution and the Fate of Humankind*. Cambridge: Cambridge University Press.
Daly, M. (1978) *Gyn/Ecology: the Meta-ethics of Radical Feminism*. Boston: Beacon Press.
Garnaut, R. (2008) *Garnaut Climate Change Review: Final Report*. Melbourne: Cambridge University Press.
Goodpaster, K. E. (1978) "On Being Morally Considerable." *Journal of Philosophy*, 75, 6, 308-325.
Griffin, S. (1978) *Woman and Nature*. New York: Harper Colophon.
Jehne, W. (2007) "The biology of global warming and its profitable mitigation." *CSIRO Sustainability Network Update*, Feb.
King, Y. (1990) "Healing the Wounds: Feminism, Ecology and the Nature/Culture Dualism," in I. Diamond and G. F. Orenstein (eds.), *Reweaving the World*. San Francisco: Sierra Club Books, pp. 106-121.
Lloyd, G. (1984) *The Man of Reason*, London: Routledge.
Mathews, F. (2008) "Thinking from Within the Calyx of Nature." *Environmental Values*, 17, 1, 41-65.
Mathews, F. (2006) "Beyond Modernity and Tradition: towards a Third Way for Development?" *Ethics and the Environment*, 11, 2, 85-114.
Mathews, F. (2003) *For Love of Matter: towards a Contemporary Panpsychism*. Albany NY: SUNY Press.
Mackey, B. G., Keith, H., Berry, S. L., and Lindenmayer, D. B. (2008) *Green Carbon: the Role of Natural Forests in Carbon Storage*. Acton ACT: ANU epress.
Mshigeni, K. (2001) in *Earth Times*, San Diego. Retrieved in January 2009 from <www.sdearthtimes.com/et0101/et0101s7.html>.
Naess, A. (1973) "The Shallow and the Deep, Long-range Ecology Movement: A Summary." *Inquiry*, 16, 95-100.
Passmore, J. (1974) *Man's Responsibility for Nature*. London: Duckworth.
Plumwood, V. (1993) *Feminism and the Mastery of Nature*. London: Routledge.
Rodman, J. (1983) "Four Forms of Ecological Consciousness Reconsidered" in D. Scherer and T. Attig (eds.), *Ethics and the Environment*. Englewood Cliffs, NJ: Prentice Hall, 82-92.
Rolston III, H. (1975) "Is There an Ecological Ethic?" *Ethics*, 85, 93-109.
Rose, A. L. (2007) "Biosynergy: the Synergy of Life". Retrieved in January 2009 from <www.biosynergy.org/docs/encyclopedia-article.pdf>.
Routley, R. (1973) "Is There a Need for a New, an Environmental Ethic." Varna (Bulgaria): *Proceedings of the XVth Congress of Philosophy*, pp. 205-210.
Routley, V. and Routley, R. (1980) "Human Chauvinism and Environmental Ethics," in D. Mannison, M. McRobbie and R. Routley (eds.), *Environmental Philosophy*. Canberra: Research School of Social Sciences, Australian National University, pp. 96-189.
Ruether, R. R. (1975) *New Woman/New Earth*. Minneapolis, MN: Seabury Press.

Sagoff, M. (1984) "Animal Liberation and Environmental Ethics: Bad Marriage, Quick Divorce," *Osgoode Hall Law Journal*, 22, pp. 297-307.
Saunders, D. (2000) "Learning From Nature." *Living Lightly* supplement of *Positive News,* Spring. Retrieved in January 2009 from <www.ceu.org.uk/proposal/appendix/positivenews.ze.doc>.
Taylor, P. W. (1981) "The Ethics of Respect for Nature." *Environmental Ethics*, 3, 3, pp. 197-218.

4

Ethical and Prudential Responsibilities, Culture and Climate Change

Thomas Heyd

In the face of the increasing certainty about the powerful impact of climate change, a premature sense of defeat, with a consequent tendency to not do anything much regarding this global phenomenon, can easily set in. As I will point out, the view that we really need not do much about climate change may even *appear to be* ethically justifiable. It is, however, of the greatest importance to avoid that defeatism and avoid becoming a self-fulfilling prophecy. In this chapter, I seek to address this quandary. First, I argue that, even if it may appear that we cannot effectively act on our responsibilities regarding climate change directly, because of certain conditions prevalent in our societies, we may still have a responsibility to act on another level. Second, I propose that, even if direct action upon this phenomenon may seem ineffectual, we may still be able to address climate change by taking note of the cultural frameworks that enable and constrain our actions. I illustrate this point with regard to the role of diverse conceptions of natural forces. I conclude that the consideration of the cultural dimensions of responses to natural forces likely is of key importance in the development of the sort of coping and resilience that are needed to satisfactorily confront climate change.

Ethical and Prudential Responsibilities and the Relevance of Culture

> *Warming of the climate system is unequivocal, as is now evident from observations of increases in global average air and ocean temperatures, widespread melting of snow and ice and rising global average sea level.*
>
> – IPCC 2007, 30

The last report of the Intergovernmental Panel on Climate Change (IPCC 2007) has strongly supported the view that climate change processes are accelerating and that future as well as present generations are going to experience very significant increases in the severity of disastrous, and possibly catastrophic, natural phenomena. It is foreseeable that human populations, especially those least well off, will be most affected, and that human-induced species extinctions will accelerate.

Though climate change may seem to be something new, there has been significant climate variability at various points during historical periods (see, e.g., Fagan 2000). In prehistory, moreover, modern human beings lived through repeated periods during which they experienced rather sudden climate change (see Burroughs 2005). We also know that some groups of human beings better weathered such periods than others. This should make us hopeful that human beings can indeed address this situation if they take appropriate measures.

It is evident that the potential harm to human and non-human beings arising from climate change calls for consideration of our ethical responsibilities.[1] These responsibilities are based on a range of principles, beginning with the basic notion that actors bear responsibility for the harm they produce (expressed often in the "polluter pays" principle) to the view that those who can prevent considerable harm from occurring, ethically, ought to prevent it (cp. Singer 1993). Moreover, given the very considerable extent of the expected harmful effects for human beings resultant from climate change, we can also assume that a strong case can be made for responsibilities from the perspective of prudence. This means that we not only have responsibilities to act insofar as we are individual moral agents, but also as persons with cares for self who are enmeshed in the global society. Our implication in global processes, in turn, translates into additional roles as social actors, in the sense that, both for ethical and prudential reasons, we ought to act for the common well-being in which we all share. This role as social actor comes into play insofar as we fulfill roles as elected representatives of society, functionaries of governments, decision-makers in private enterprise, or simply as members of civil society.

Our responsibilities to act with regard to climate change can be understood in terms of prevention, mitigation, or adaptation. Since it is too late for outright prevention, this leaves mitigation and adaptation, and given the pace at which climate is changing and the scale of its impacts, it means that the responsibility to act is urgent. In summary, I am assuming that it is quite clear that climate change constitutes an urgent moral and prudential problem, such that the need for action is beyond reasonable debate, even if the scope of action required, the distribution and source of responsibilities, or the assignment of benefits and costs, may be worth considering in further detail (see Gardiner 2004, Garvey 2008).

This raises the question of whether our responsibilities to act can be appropriately discharged. Famously, "ought implies can." The idea is that obligation ceases if one cannot discharge it. For instance, saving someone on the Moon may be impossible for me, even if it were my responsibility, and hence, technically, my responsibility to act, whatever its grounding, would cease—though I might feel regret at my failure to act. So, *can* we act on our responsibilities to mitigate climate change, as well as adapt our societies for the impacts that are already on their way?

On the one hand, the answer is a resounding *yes*, as the IPCC Assessment Reports and other studies have persuasively argued. We do have the material resources and know-how, in the world community, to effectively address these responsibilities, even if there is debate on the extent of the changes that this would impose on our present lifestyles. As laid out in various international reports, such as are published by the IPCC, greenhouse gas emissions need to decrease rapidly, reaching a small fraction of their present levels within a few decades. Basically, it means that we need to move to a carbon-free way of living as quickly as possible. This will doubtlessly mean important changes in our ways of working and living, but various studies suggest that we could indeed achieve this goal if we are really committed to it.

On the other hand, the answer seems to be *no*, at least when given the present set of beliefs, values, priorities, practices, and material culture prevalent in our societies. So, are we condemned to the assumption of defeat before we even try? There are some reasons to be pessimistic about the readiness, and even the capacity, of people to act and of society's decision-makers to facilitate the deep, and possibly drastic, changes to our lifestyles needed to address climate change in an effective way. The record is not encouraging. To date, there has been very little action to mitigate, and to adapt to, climate change. This is the case even though it has been known, since 1896, that increases in carbon dioxide could raise the level of the planet's average temperature, and that, since the 1930s, there has been awareness of warming in North America (Weart/American Institute of Physics 2003-2007). In 1988, the World Meteorological Association, together with the United Nations Development Program, created the IPCC to assess the impacts of climate change and to propose options to respond to it. This has been followed by various efforts at international, national, regional, and local levels to develop international law and regulation to address climate change, including the Kyoto Protocol.

Nonetheless, despite international agreements, there still are national governments of countries, such as Canada, that are major contributors to the generation of greenhouse gases, but only give lip service to the notion that something needs to be done urgently about climate change. (Canada has even ratified the Kyoto Protocol, but the present government has so far refused to honor its commitment.) When something is done, finally, it generally is too little carried out too slowly. The same is true with regard to most individuals. So, if, given our present economic/social/political context, we *could not* make a real difference, either as individuals or as social actors dependent on ineffectual institutions, while trying to discharge our moral and prudential responsibilities regarding climate change, these responsibilities would, arguably, be held in abeyance or suspended. Not doing anything would be justified, or at least not be considered a cause for blame. So where are we? Is it still possible to act on one's responsibilities with regard to climate change? Or are we condemned to inactively await the fates that will befall us? This stark situation calls for further elucidation.

There are reasons to believe that standard approaches to adaptation, which rely on the application of technical, managerial, or scientific know-how, by themselves are not sufficient to address the realities of climate change. Changes in the way that we approach climate change depend on people and, more broadly, on the ensembles of beliefs, values, habits, practices, and material culture that enable and constrain thought and action. We may call these ensembles "cultural frames" or "cultural frameworks."[2] The effectiveness for generating change in behaviors and in society, and the effectiveness of new information, new technical devices, or new management schemes, depends on their mediation through the cultural frameworks of the people whose active cooperation such innovations generally require. If the innovation does not take into account the extant cultural frameworks of those whose participation it requires, it may end up being rejected or neglected, with consequent loss in effectiveness.

The role of cultural frameworks is evident even with regard to issues concerning which knowledge seems readily available to the public concerned. For example, despite repeated warnings about increasing frequencies of high level floods, people continue moving to, and building conventional housing in, flood plain areas in Vancouver, Canada (in the municipalities of Richmond and Delta). As a consequence of climate change, these parts of the city, however, are foreseeably going to be inundated, both as a result of higher levels of the Fraser River, which flows through a great part of the metropolitan area, and due to rises in sea level. In Spain, similarly, countless new, intensely water-demanding, golf course developments are being planned and undertaken even though it is expected that climate change will increase the frequency and severity of recurring droughts in that country. In fact, accelerated desertification is expected to put a great strain on water supplies even for basic needs (drinking water supplies and traditional irrigation agriculture).

The importance, for adaptive behavioral patterns, of the array of beliefs, values, priorities, practices, and material culture that constitute the cultural frames that guide human action is particularly striking and instructive in relation to responses to severe natural events. Even when such events directly affect human populations, people all too often continue with behavioral patterns that seem poorly adapted to the hazards of the area. It is noteworthy that, *even after* important disasters, people would choose to ignore—or even embrace—the inherent risk of locations prone to flooding, earthquakes, volcanic eruptions, or landslides. A case in point is the rebuilding of towns and villages along shallow sea shores after the devastating 2004 tsunami in Sumatra. Notably, tourism-oriented towns are rebuilt in places where mangroves, which constitute a sort of natural shield against storm surges and tsunamis, formerly grew. Another case is the continued occupation of riparian areas in Central Europe that in recent years repeatedly suffered severe flood damage.

When we try to understand these sorts of reactions, the deeply rooted assumption often seems to be that such severe events are rare and can be ignored, even while the probabilities of the occurrence of such incidents are changing. The explanation for the lack of adaptation in such cases may, moreover, be sought in a diversity of conditioning factors, including particular social, geographical, and economic circumstances. People often end up rebuilding in the same places and exposing themselves to the same risks as before such drastic events, due to economic and social marginalization, or imposition by authoritarian regimes, or short term profit-oriented economic incentives, or simply the legal need to maintain their presence on property that they depend on (Leroy 2005). The failure to deploy suitable coping strategies in the rich countries of this planet, even when people *have* sufficient scientific information, technical expertise, and managerial know-how to change course, supports the view that the role of cultural frameworks is of fundamental importance. In other words, the absence of more appropriate disaster prevention strategies can be explained, at least up to a point, by the set of beliefs, values, priorities, practices, habits, and material culture, both enabling and constraining people's responses to potential environmental hazards.

We return to the question of whether we can really act on our ethical and prudential responsibilities with regard to climate change if it appears to be ineffectual to do so. Notably, action may seem ineffectual in generating the necessary change because of the prevalent cultural framework of our societies, which, among other things, favors short-term benefits over long-term risk reduction, as well as "management of risk" and a "fix-it-after-the-fact" approach in preference to prevention, mitigation, and adaptation. One response to the question of whether we *can* act on our responsibilities is that, even if there is no effective way of *directly* acting on them, there may be options for effectively acting on them in an *indirect* way. This can be conceived somewhat on the pattern discussed by Harry Frankfurt (1971) concerning the relation between first and second order desires (or "volitions").

According to Frankfurt, we may, for instance, not have a desire presently to bicycle to work or to exercise daily, but may *want to have that desire*: we may envy people who do have those desires, for example. And, if we desire enough to acquire such (first order) desires then we may go as far as changing the circumstances that we are in, such that we end up acquiring those (first order) desires that we want to have. On this account, the desire to acquire the first order desires is a second order desire. My proposal is that, just as we may have second order desires to bring about certain first order desires, we, similarly, may perceive certain *second order responsibilities* to bring about the conditions for exercising our first order responsibilities. So, even if we cannot act on our first order responsibilities with regard to climate change, because of the circumstances in our societies, we may still be able to act on our second order responsibilities with the aim of bringing about conditions such that acting on our first order responsibilities may eventually become effective.

Given that the major impediment in acting on mitigation and adaptation regarding climate change seems to have its roots in the prevalent cultural framework, acting on our second order responsibilities may mean *seeking ways to transform this framework*. If cultural frameworks are made up of ensembles of beliefs, values, priorities, habits, practices, and the like, in a context of a particular material culture, then transforming those frameworks in ways such that they may be more suitable to address climate change may mean finding, and applying, key points of leverage. Broadly speaking, this means pursuing and supporting research, as well as engagement at the personal and institutional level, in projects that can bring about the changes needed.

Determining those points of leverage for transforming cultural frameworks partially is an empirical issue, to be settled with the help of social and psychological sciences. Nonetheless, insofar as changing ways of living depends on a *commitment to a different world view*, it also concerns individuals and communities understood as *agents* who have the capacity and the desire to maintain their sense of *autonomy*. As such, it is a matter to be explored in terms of the coherence of conceptions of the world with personal and community autonomy, the support that those conceptions give to personal and community-wide value systems, the priorities that those value systems call for or allow, the habits and practices that make sense given those contexts, the material culture that these conceptions require, and so on. These are matters that call for consideration from the perspective of philosophy, history, literature, the visual arts, and even architecture, urban design, and so on, in addition to the light that the social sciences may throw upon them.

Our conception of nature is a key element in our cultural framework when we envisage interaction with the natural environment since it interacts with our values, priorities, and practices, in relation to that environment. Since present climate change, even if co-produced by human activities, is a phenomenon resultant from the interaction of various natural phenomena (including wind patterns, ocean currents, and absorption of heat in the atmosphere), it is relevant to consider how, in general, our conception of natural forces impacts our readiness to cope with severe natural events, and in particular, our readiness to bring about mitigation and adaptation regarding climate change.

Culture, Conceptions of Natural Forces, and Coping

The notion of culture, and how to understand it insofar as it mediates our relationship to the environments in which we live, is itself debated (see, e.g., Ingold 1994). Certainly, we should not think of cultures as neat, homogeneous, isolatable units that can be apportioned to discrete human groups. The cultural frameworks that enable or constrain people's thought, action, habits, and practices have their foundation in particular social-historical contexts, but should be conceived as dynamic, subject to constant transformation, and in regular interaction with the frameworks of people from other social-historical *milieux*. This is especially true today, given the inter-relationship of human populations in our increasingly globalizing world. Any set of values, beliefs, and practices is, moreover, mediated by power relations, and not simply the result of adaptation to objective conditions of the natural environment. We may assume, nevertheless, that people develop particular cultural frameworks in interaction with the local natural environment on which they depend, in an effort to satisfy particularized needs and desires in the face of specific obstacles and opportunities. As such, these frameworks condition the degree to which people may be more or less ready to cope with powerful natural events and processes present in their environments.

The impact of differences in culture with respect to mitigation and adaptation may be illustrated even with an example from a relatively homogeneous Western context.[3] We may consider, for instance, that Canada and Scandinavian countries, such as Sweden, probably have access to equivalent levels of expertise with regard to technologies that can facilitate carbon-neutrality, and enjoy similar types of democratic institutions and overall comparable levels of economic well-being. Nonetheless, while situated in relatively similar climates, Scandinavian countries are moving ahead much faster than Canada in making their economies independent from oil and gas. Sweden, for example, seems well on its way toward achieving this goal by 2020 (see Giddens 2006, 145; Vidal 2006, 16). The differences in policies certainly can be analyzed in terms of availability of resources and historical-political and economic factors, but they likely are rooted in something more fundamental, such as differences in the habits of taking prudential, long-term oriented measures, along with differences in the corresponding attitudes, values, belief systems, and structure of institutions.[4]

Differences in cultural frameworks, furthermore, may provide a plausible explanation for differences in precautionary measures taken by diverse groups of people with regard to events experienced as "disasters." For some populations, when such events occur, they remain anchored in their cultural memory, at least for the time span of one lifetime, while for other populations, only a few kilometers away, who have undergone similar experiences, this is not so. We observe this, for example, with regard to awareness of the signs of impending tsunamis and volcanic eruptions among populations living in Papua New Guinea and the Solomon Islands (Davies 2002, 37–38). After a disaster, some populations will even be willing to relocate their villages permanently, while others nearby refuse to or return after a short absence (Davies 2002, 39–40). In some cultural contexts, disastrous natural phenomena will lead to the creation of myths and the establishment of taboos about occupying certain areas of the land, while not in others (see, e.g., Lowe et al. 2002, 138, on such taboos in New Zealand). These differences may be explicable by differences in cultural frameworks mediating the conceptualizations of such events, even if other factors, such as differing levels of social cohesion, historical memory, and so on, also play a role. In general, such cases suggest that cultural frameworks are fundamental factors in human responses to natural changes by the way in which they mediate understanding of, and coping with, those phenomena. Applied to our contemporary situation, this has some interesting implications, which can only be sketched here, however.

The examples described suggest that it may be valuable to reflect on how our cultural frameworks structure the conceptualization of the natural forces that predominate in our Westernized global society, and the implications that such conceptions have for the readiness to confront disastrous natural phenomena such as are to be expected in relation to climate change. It also means considering how those conceptions of nature interact with the beliefs, values, priorities, practices, and material culture that shape peoples' social and economic activities. Given that some of our present problems are the result of viewing disruptive natural events as merely temporary or even only momentary obstacles, of which it is supposed that they can eventually be controlled, it may be interesting to consider alternatives, such as the idea that natural phenomena can exhibit a kind of self-organization or agency that may only be controllable or predictable in a very limited way (Homer-Dixon 2006; Heyd 2005, Heyd 2007, and Heyd forthcoming). The practical consequences of such considerations in relation to climate change are far reaching. One straightforward practical conclusion, for example, may be to give natural entities and forces space and time for their expression. This sort of idea has already been adopted by some environmental managers who argue for the rehabilitation of deltas and polders as riparian flood retention areas, or the restoration of mangroves and coastal forests in threatened coastal areas (Ledoux et al. 2005).

Conclusion

I am assuming that, as ethically and prudentially concerned individuals and social actors, we have responsibilities to limit the harm to human and non-human beings due to climate change. While the challenges posed by mitigation and adaptation regarding climate change certainly call for the development of ingenuity and technical-scientific and managerial expertise in order to come up with appropriate physical and socio-economic modifications to our environments, they will also require fundamental changes in people's ways of living.

I have proposed that, even if—due to the present conditions of our societies—there may be limitations in our ability to directly act on our responsibilities to bring about mitigation and adaptation regarding climate change in a direct way, there is another order of responsibilities on which we may act. This centrally involves the transformation of prevailing cultural frameworks, which, among other things, means reflection on our conceptualizations of natural forces.

The anthropologist Julie Cruikshank encapsulates the importance of the wider cultural dimension in her claim that "our human ability to come to terms with global environmental problems will depend as much on human values as on scientific expertise ..." (Cruikshank 2001, 390). We can hope that by reflecting on, and transforming the conceptions, values, practices, priorities, and material culture that make up the prevailing cultural framework, we will be enabled to develop the robust coping behaviors and resilience that we need to effectively face the challenges posed by climate change.

Postscript

The meeting organized by the Toda Institute has proven that dialogue about alternative ways of conceiving of our relation to the natural world can be very effective in generating a spirit of collaboration and a readiness to think afresh about our values, and about our ethical and prudential responsibilities regarding climate change. I congratulate the organizers on this successful event and hope that it may lead to many further initiatives to renew our readiness to take on our responsibilities regarding climate change.

Notes

1. I do not intend to prejudge the outcome of philosophical discussion on the foundations of ethical responsibility regarding climate change, but assume that most reasonable stances, coherent with common morality, will lead to the conclusion that, given the urgency of the situation and the extent of the harms in view according to the most recent scientific findings, some action is required. E.g., Gardiner 2006 as well as Garvey 2008 survey the diverse arguments in circulation.
2. The supposition that cultural frameworks are of key importance in bringing about the change that is needed with regard to climate change constitutes a relatively new

research approach. There are a growing number of research programs that together are triangulating on it, however. See, e.g., Heyd (2007), Gifford (in press), Crate (2008), Lahsen (2008), deMenocal (2001).

3. Though many commentators speak of "the West" as if it were one single cultural unit, the actual differences between countries and even regions can be rather stark. Compare the difference between German and Spanish cultures, or even among Basque, Castilian, and Catalan cultures, all found within the country Spain.
4. The cultural differences may be of an even more fundamental sort, of course, perhaps affecting the most basic conceptions of individuals as selves, of society, and of natural phenomena.

References

Burroughs, W. J. (2005) *Climate Change in Prehistory: The End of the Reign of Chaos.* New York, USA: Cambridge University Press.

Crate, Susan A. (2008) "Gone the Bull of Winter? Grappling with the Cultural Implications of and Anthropology's Role(s) in Global Climate Change." *Current Anthropology*, (August), *49* (4), 569-95.

Cruikshank, J. (2001) "Glaciers and climate change: Perspectives from oral tradition." *Arctic*, *54* (4), 377-93.

deMenocal, Peter B. (2001) "Cultural Responses to Climate Change During the Late Holocene." *Science*, (27 April) 292 (5517), 667-673.

Davies, H. (2002) *Tsunamis and the Coastal Communities of Papua New Guinea, in Torrence, R. and Grattan, J. (eds.) Natural Disasters and Cultural Change.* London and New York: Routledge, 28–32.

Fagan, B. (2000) *The Little Ice Age: How Climate Made History 1300-1850.* New York: Basic Books.

Frankfurt, H. G. (1971) "Freedom of the Will and the Concept of a Person." *The Journal of Philosophy*, 68 (1), 5-20.

Gardiner, S.M., (2004) "Ethics and Global Climate Change." *Ethics*, 114, pp. 555–600.

Giddens, A. (2006) *Debating the Social Model: Thoughts and Suggestions. In The Hampton Court Agenda: a Social Model for Europe.* London: Policy Network, pp. 95-150.

Gifford, Robert (In press) "Psychology's Essential Role in Alleviating the Impacts of Climate Change," in Mary Gick (ed.), special issue of *Canadian Psychology/psychologie canadienne.*

Heyd, T. (ed.) (2005) *Recognizing the Autonomy of Nature: Theory and Practice.* New York: Columbia University Press.

Heyd, T. (2007) *Encountering Nature: Toward an Environmental Culture*. Ashgate, Aldershot, U.K.

Heyd, T. (Forthcoming). "Philosophy and Climate Change." *Human Ecology Review.* Homer-Dixon, T. (2006) *The Upside of Down: Catastrophe, Creativity and the Renewal of Civilization.* Toronto: Knopf.

Ingold, T. (1994) *Introduction to Culture,* in T. Ingold (ed.), *Companion Encyclopedia of Anthropology, Humanity, Culture and Social Life*. London: Routledge.

IPCC (Intergovernmental Panel on Climate Change) (2007) *Climate Change 2007: Synthesis Report*. IPCC: retrieved on 4 January 2009 from <http://www.ipcc.ch/ipccreports/ar4syr.htm>.

Jamieson, D. (2001) *Climate Change and Global Environmental Justice*, in P. Edwards and C. Miller (eds.), *Changing the Atmosphere: Expert Knowledge and Global Environmental Governance*. Cambridge, MA: MIT Press, pp. 287-307.

Lahsen, Myanna (2008) "Commentary 'Gone the Bull of Winter? Grappling with the Cultural Implications of and Anthropology's Role(s) in Global Climate Change.'" *Current Anthropology*, (August), *49* (4): 569-95, pp. 587-88.

Ledoux, L., S. Cornell, T. O'Riordan, R. Harvey and L. Banyard (2005) "Towards sustainable flood and coastal management: identifying drivers of, and obstacles to, managed realignment." *Land Use Policy*, 22, 129-144.

Leroy, S. (2005) *Rapid Environmental changes and civilisation collapse: can we learn from them? Rapid Landscape Change and Human Response in the Arctic and Sub-Arctic Conference*, 15-17 June 2005, Whitehorse, Yukon, Canada.

Lowe, D. J., R. M. Newnham and J. D. McCraw (2002) *Volcanism and early Maori society in New Zealand*, in R. Torrence and J. Grattan, *Natural disasters and cultural change*. London and New York: Routledge, pp. 126-161.

Singer, Peter (1993) *Practical Ethics*. Cambridge: Cambridge University Press.

Vidal, John (2006) "Sweden plans to be world's first oil-free economy," Top stories section, *The Guardian* (8 February 2006): 16; retrieved on 4 January 2009 from <http://www.guardian.co.uk/environment/2006/feb/08/frontpagenews.oilandpetrol>.

Weart, S. and American Institute of Physics (2003-2007) *The Discovery of Global Warming, Introduction: A Hyperlinked History of Climate Change Science. Center for History of Physics*. American Institute of Physics: retrieved on 30 March 2008 from <http://www.aip.org/history/climate/summary.htm>.

Part II

Governance

5

Closing the Boxes, Enlarging the Circles: Toward a New Paradigm of Global Governance and Economy

Sheila D. Collins

There is a scientific consensus that the earth is at a dangerous tipping point, yet our political and economic institutions lag far behind the need to act quickly and decisively on climate change. This chapter seeks to answer the question of why, despite what modern science has known about the world and what climate scientists have been warning us about for decades, do our political and economic institutions continue to be so recalcitrant? How might those concerned about climate change begin to affect our decision-making apparatuses? My answer to this problematic draws on two strands of thinking about the sources of the problem: the work of Debeir, Deléage and Hémery (1991) in tracing the central role of energy systems in human development and in the erection of political/economic regimes that resist change; and the application of Thomas Kuhn's theory of paradigm shift in science to the realm of political ethics by cognitive scientist George Lakoff. I then use Lakoff's theory about how the mind works to decode the Enlightenment paradigm that has shaped the nation-state system and the capitalist economy in order to show how its conceptual distortions have impeded us from moving toward a new ecological paradigm based on how the earth system actually works. Finally, I develop a schematic framework that contrasts the Enlightenment paradigm with the ecological paradigm and suggest some ways of working to bring it into sharper focus.

> *You have noticed that everything an Indian does*
> *is in a circle, and that is because the Power of the*
> *World always works in circles, and everything tries*
> *to be round. . . .*
>
> – Hehak Sapa (Black Elk)[1]

> *The country was made without lines of demarcation, and it is no man's business to divide it . . . . The earth and myself are of one mind. The measure of the land and the measure of our bodies are the same.*
>
> – Chief Joseph, Nez Pierce[2]

> *God and his [man's] reason commanded him to subdue the earth, i.e. improve it for the benefit of life, and therein lay out something upon it that was his own, his labour. He that in obedience to this command of God, subdued, tilled and sowed any part of it, thereby annexed to it something that was his property, which another had no title to, nor could without injury take from him.*
>
> – John Locke[3]

Introduction

Despite scientific consensus that the earth is reaching a dangerous tipping point, the institutional systems through which decisions are made about how to manage the earth's resources lag far behind the need to act quickly and decisively (Hansen 2008, IPCC 2007).[4] The international community has failed to meet the Kyoto targets for reducing greenhouse gas emissions in the years since they were adopted, and the United States has failed to commit to binding targets at all.

Why do our political and economic institutions continue to be so recalcitrant? How might we begin to impact the decision making process? That is the conundrum this chapter seeks to illuminate.

Sources of the Problem

My attempt to answer this problematic draws on three strands of thinking about the sources of the problem. The first is a thesis proposed by Debeir, Deléage and Hémery (1991) in their evaluation of the role of energy in historical development. The centrality of energy to an understanding of world history has curiously been neglected by historians, yet it is the fulcrum of all human history because it is what makes the reproduction of human life, all work, and thus civilization possible. "Energy is present all along the chain of causes and effects from which human evolution derives" (Débeir, Deléage & Hémery 1991, 9).[5] They argue that in every historical epoch one form of energy predominates over others. History can thus be classified on the basis of the energy systems that predominate in certain epochs—the "energy system" being defined as the predominant energy source, the system of converters that turns that source into useful work, and the social organization that monopolizes and coordinates its use. Energy systems give rise to classes that become powerful on the basis of their ability to harness and monopolize energy sources and their converters. Throughout pre-history, human labor power was the predominant form of energy, but to harness it on a scale necessary to produce more than the simple reproduction of the laborer, and thus the beginnings of civilization, required both favorable geography and a coercive state. Antiquity was characterized by the coordination and mechanization of human labor power by a centralized political-religious state; the medieval age by the feudal system in which secular lords monopolized the hydraulic system (the water mill, later aided by the windmill); and the early industrial age by the factory system that enabled the rising capitalist class to harness thermal energy by means of the steam engine (Débeir, Deléage & Hémery 1991, 7). During the twentieth century, the fossil fuel system predominated and elites tied to the fossil fuel industry exercised political power (Gelbspan 2004).

Because those who control the monopolization of an energy source enjoy tremendous power, these ruling elites are unwilling to give it up and will continue to exploit the useful application of that form of energy through the invention of more sophisticated machinery until it has reached its saturation point or there is a societal collapse. Even if other forms of power are discovered, they will be underutilized until the previous two conditions are met. Those who control an energy system are able to prolong its life for two reasons. First, energy systems are highly adaptable to technical innovation, to sudden spurts in energy demand, to new tools and relations of production, and to the development of extended areas for extraction. Second, those who control the system are often able to recycle in different guises energy technologies that have fallen into disuse in order to prolong the life cycle of the preferred energy source. Thus, slavery, which had fallen into disuse in medieval Europe, returned as European capitalism surged across the Atlantic. "The human engine was the ultimate solution to the Americas' prolonged energy shortage" (Débeir, Deléage & Hémery 1991, 11). Similarly, as we now approach the period of peak oil, we see the return of three earlier outmoded systems of energy: slave labor, indentured servitude, and sweatshops, which all serve to prolong the life-cycle of the fossil fuel system.

As population growth increases, energy must be applied more intensively to drive up labor productivity. Until the invention of electricity, demand for energy was kept within limits by the number of daylight hours available for work. Energy forms suitable for artificial lighting were developed so that labor and new machines could be used more intensively. The revolution in telecommunications and transportation during the last thirty years has made possible a global assembly line that has intensified the use of energy and labor as it both serves an expanded population and feeds its growth.

This explanation of energy's role in history helps to explain the significance of the role of energy-related elites in the governing structures of the most powerful countries during the twentieth century, especially that of the post-WWII United States.[6] It explains the British and American governments' role in the overthrow of Mohammed Mossadegh in Iran, the strange alliance of several administrations with the Saudi royal family, the Bush administration's desire to wage an illegal war for oil in the Middle East, its suppression of the findings of its own climate scientists, its attempt to discredit the Intergovernmental Panel on Climate Change (IPCC), and its opening of previously off-limits and more expensive areas to find oil, coal, and gas, while discrediting the viability of renewable energy sources as "too expensive" (Gelbspan 2004, Johnson 2004, Klare 2004). The rise of the new "petro states" that are now challenging Western-dominated energy elites is but the latest chapter in the ongoing saga of the prolongation of energy systems.

The second strand of thinking about the sources of the problem of political and economic inertia on meeting the challenge of climate change comes from Thomas Kuhn's work on paradigm change in science (1996) and its application by cognitive scientist George Lakoff to the realm of politics and ethics. Kuhn showed that the normal workings of science are deeply influenced by received beliefs or paradigms that are slow to change, can last for centuries, and are only transformed when they meet with a series of anomalies for which previously held theories have no explanatory power. Science—based on empirical proof—has a method for correcting itself once these anomalies become recognized widely enough. Neither economics nor politics is based on such a method, although both professions pretend to be scientific modalities. The nation-state system and the capitalist economy emerged at about the same time as the emergence of Enlightenment philosophy, but these two decision-making systems never got beyond the Enlightenment paradigm with which they began. Thus, what we have operating at the center of the two institutional systems through which all of the major decisions about how to order our collective life and manage our energy systems are made is a paradigm of the world that has become dysfunctional, if not pathological, as we attempt to meet the crises posed by climate change

(Capra 1997). Ecological philosopher Thomas Berry has said that such a mode of thinking represents a state of disturbance that "exists at a more basic level of consciousness and on a greater order of magnitude than we dare admit to ourselves ..."

> This unprecedented pathology is not merely in those more immediate forms of economic activity that have done such damage; it is even more deeply imbedded in our cultural traditions, in our religious traditions, in our very language, in our entire value system (Berry 1988, 10).

We find an explanation as to why these two systems have been unable to break out of this outmoded paradigm in the work of the cognitive scientist George Lakoff (Lakoff 2008). Politics, says Lakoff, is "centrally about ideas, actions, perceptions, policies, and communication, all of which require an understanding of the mind" (Lakoff 2007, 70). New discoveries in cognitive science have uncovered the ways in which the brain comes to formulate the sets of ideas and beliefs that Kuhn describes as a paradigm, as well as how they affect human and animal behaviors in everyday life as well as in science, politics, and economics.

According to Lakoff, cognitive neuroscience has given us a picture of the brain as a deeply embodied set of processes that extends throughout the human body. Ninety-eight percent of what the brain does is at the unconscious level. The ideas that the brain comes up with depend on the peculiarities of human anatomy as well as on our social interactions (Lakoff 2008, 10; 2007, 70). Lakoff contrasts this image of the deeply embodied brain with the image of the brain—and its product, reason—that emerged from Enlightenment philosophy and that persists today in the ways we commonly think about our relation to the world and sociopolitical life. In Enlightenment thought, our ability to reason is what makes us human. For Enlightenment philosophers, reason is logical (it leads from facts to correct conclusions), unemotional, value-neutral, interest-based (it serves one's purposes), and literal—it perfectly fits the logic of the objective world (Lakoff 2008, 7).

In contrast to this view, neuroscience has shown that "reason is mostly unconscious and physical. It requires emotion and operates through image schemas, conceptual frames, and prototype structures that form themselves into metaphors when two different parts of the brain are regularly activated" (Lakoff 2007, 70). In turn, metaphors combine to build more complex narratives. These metaphors are both inside the body—"in the very building blocks of our brains" as well as in our cultures and languages (Lakoff 2008, 21; 1980). The institutions we create, in turn, reinforce those metaphors and cement the neural circuitry through which they work. Frames, a shorthand way of describing this process of unconscious thought, are important, because they help us to makes sense of "facts." But frames can also exclude facts. They can literally cause us not to see—not even to imagine—that which the conceptual frame does not include in its repertoire.

In its ability to describe the nature of the universe, modern science has gone beyond the seventeenth century frames with which it began. However, the nation-state system and the capitalist economy which dominate much of the world today continue to operate through Enlightenment frames. The values such metaphors and narratives enshrine and the practices that have resulted have brought us to the brink of climate havoc. Enlightenment thought and practice have undeniably brought us much that is good, both in the realm of scientific and technological advances as well as in the realm of political development. The idea that there are certain human rights that ought to be inalienable and that human beings have the ability and right to self-governance are Enlightenment ideas that stand the test of time. But these ideas are based on a deeper set of metaphors and narratives that give a particular shape to the way in which they are interpreted. The unreflective adherence to Enlightenment epistemology in the areas of governance and economy—the failure to examine and debate the merits of this deeper structure of thought—have brought us to the brink of both climate and economic crisis. Yet these dysfunctional frames are stubbornly adhered to because the institutions, which embody them, continue—through the system of normative rules, material rewards, and punishments they give out—to reinforce the unconscious neural synapses of the Old Order. Only after we bring them to consciousness and decode how they work can we fashion new frames more consonant with what we now know of the nature of the earth system and with new socioeconomic/political relationships that embody this changed conceptual world.

A central metaphor for the nation, Lakoff argues, is that of family—a metaphor that arises from earliest childhood in the interaction between parents and children (Lakoff 2008, 73). The nation as "family" can be extended to many other kinds of institutions, such as corporations, armies, organized religions, etc. (Lakoff 2008, 82-87). Lakoff then discusses two primary models and a few basic variations of the Nation as Family metaphor that manifest themselves in American politics. One of these is the patriarchal family. "The family needs a strict father because there is evil in the world from which he has to protect them ...

and competition in the world, and he has to win those competitions to support the family—and Mommy can't do it" (Lakoff 2008, 78). We see echoes of this model in the National Security Strategy of the Bush administration and, indeed, in the foreign policies of many nation-states around the world. Lakoff contrasts the strict father model with what he calls the "Nurturant Parent" model:

> Nurturance is empathy, responsibility for oneself and others … Nurturance requires setting limits, and explaining them. It requires mutual respect—a parent's respect for children and respect for parents by children must be earned by how parents behave … The job of parents is protection and empowerment of their children, and a dedication to community life, where people care about and take care of each other … When mapped onto the nation, the result is the progressive politics of protection, empowerment, and community (Lakoff 2008, 81).

It is safe to say that the patriarchal family model still informs much of the world's governing institutions as it does much of the world's family relations. Orienting our political institutions toward a nurturing parent model would be a step in the right direction, but even Lakoff's formulation may be too culturally and anthropocentrically bound, for the earth system teaches us that we are part of a complex community made up of many forms of organic and inorganic matter and that our fate is bound up with even the lowliest worm. The notion among indigenous peoples that the rocks are "our eldest ancestors" conveys a deeper understanding of family than Lakoff's anthropocentric formulation, but his call to examine the frames that have shaped our culture is nevertheless an important one.

Let us look at some additional metaphors that have shaped the current nation-state system itself and the capitalist market. I suggest that one of the basic metaphors is that of a box. Boxes circumscribe a small space within a larger space that is amorphous and immeasurable. They shut something in and exclude everything else. They have borders that can be measured. The connection between state and box permeates the English language, which has become the lingua franca of the modern world. We speak of a "state of mind"—a time-circumscribed snapshot of a person's consciousness or mood, and the "state of nature," as if nature were a fixed entity and not a constantly evolving system of energy exchanges. A person is said to be in a "state of depression," that is, inside a bounded region, cut off from others (Lakoff 2008, 214). To be "stateless" is to be without identity. Nation-states, then, are fixed territories with defined borders. They include citizens and residents, or "subjects," and exclude "aliens" and "foreigners," who are treated as "others" and often as "enemies." It is contestable whether this dichotomy between "us" and "them" that permeates the nation-state system is an evolutionary adaptation.[7] It is, however, incontestable that nation-states foster this kind of thinking and reinforce it through rituals, propaganda, and the brainwashing of soldiers.

Yet nation-states are fictions. They are arbitrary linear lines drawn as a result of conquest and war across a natural landscape that is full of undulations, myriad ecosystems, and language and ethnic groupings. They are "unnatural." That is why when nations demarcate their borders with walls and barbed wire fences, the walls and fences feel like they are an offense to nature—an artificial encroachment.

A second metaphor that shapes the way we think about the modern nation-state system is that of the nation as a sovereign individual. In the international arena, nations have "self-interest," which is not necessarily the interest of the collective polity. They move around in space: they "*go* to war." They act: they "*make* peace" and "*sign* treaties." They take on human characteristics. Some are "benevolent" and "friendly," others are "rogues;" some are "mature," like parents, while others are "backward" or "underdeveloped."

The idea of national sovereignty arose at about the same time as the Enlightenment breakthrough in science. The view of the universe expressed by the Cartesian-Newtonian paradigm was that of a world that stood outside of and was separate from and inferior to thinking "man" (Tarnas 1991, 287). For Enlightenment thinkers, the earth and heavens were composed of "dead matter," unconscious, impersonal, and governed by regular laws that human beings could discover and learn to manipulate. Only that which could be quantified and measured was important enough to draw humanity's attention. Geometry and mathematics were the methods for discovering and manipulating this world of objects, and since humans were the crowning glory of this new cosmology, all that could be discovered through human reason was fair game for contributing to human progress conceived in linear fashion. Descartes' immodest boast is characteristic of the hubris that was to follow: "For by them [the laws of physics] I perceived it to be possible … to discover a practical [method], by means of which, we might *render ourselves the lords and possessors of nature*" [my italics] (*Discourse*).

From the guiding metaphors of science flowed the guiding metaphors of the social and economic order (Tarnas 1991, 283; Daly & Cobb 1994). As the rational human being, constituted by reason, was the apex of the natural order, so the individual atomized person—devoid of filial associations, historical antecedents, and emotion, became the prototype of political "man." Locke's statement in his *Second Treatise* is indicative:

> To understand political power right, and derive it from its original, we must consider, what state all men are naturally in, and that is, a *state of perfect freedom* to order their actions, and dispose of their possessions and persons, as they think fit, *without asking leave, or depending upon the will of any other man* [my italics] (Locke 1690, Chapter II, Sec. 4).

Such a view of the origins of humankind was, we now recognize, a fictional narrative. Locke's conception of "natural law" might be seen as the result of his attempt to translate the deeper frame provided by the biblical story of Eden, which pervaded his culture, into the epistemology of Enlightenment science and to apply it to an explanation for the origin of the emerging liberal state and the enshrinement of the law of property.

As the natural world was composed of dead matter, so apologists for the emergent market economy would conceive of land and its resources as valueless until human labor could turn them into commodities to be exchanged on the market. This contrasted with most cultures of the world, which had limited the definition of property to personal possessions, while viewing land and its resources as a common inheritance (Polanyi 1944, Hartmann 2002, Coulter 2007, 5). In the process of turning natural resources into products, land—mere "wilderness"—could be demarcated, enclosed, and owned by those whose intelligence, initiative, or—as was more often the case—brute force enabled them to do so.

Homo economicus was conceived as an individual free to buy and sell the products of his labor from the land, constrained only by the law of supply and demand that operated without human interference, much like the regular laws of nature (Smith 1991). John Locke, the father of modern liberal political theory, held the concept of private property in such high esteem, that the collective regulation of the means for acquiring private property became the central purpose of government (Locke 1690). While Locke himself claimed that "natural law" set limits on the amount of property that could be acquired by any one individual (Locke 1690, Chap. 6, Sec. 31), the set of unconscious metaphors that made up his narrative about the origins of property won the day.

Prior to this, markets, if they existed, had been embedded in social relations. Community values dictated the way in which the gifts of nature were exchanged (Hyde 2007). Not even the mercantilist form of capitalism that had preceded market capitalism had made this separation; but when market capitalism became full blown:

> Instead of economy being embedded in social relations, social relations are embedded in the economic system ... For once the economic system is organized in separate institutions, based on specific motives and conferring a special status, society must be shaped in such a manner as to allow that system to function according to its own laws (Polanyi 2001, 60).

Thus was the origin of the idea that government and the economy were two distinct realms and that government should interfere as little as possible in the self-regulating market. According to Polanyi, this separation was an entirely new phenomenon in human history when it emerged at the end of the eighteenth century, and it represented nothing less than the complete transformation of society (Polanyi 2001, 74).

As the market economy evolved, the metaphor of empty land as useless came to justify private ownership through conquest, wars, theft, and fraud of vast areas of the world's lands and waters, as well as its mineral deposits and now even its air, its gene pool, and its artistic and intellectual creations. Since the state was now subordinated to the law of the market, its existence came to depend on the efficient operation of the market mechanism, which meant giving a central place to the value of economic growth, measured as an increase in the rate of profit and in the quantity of GDP, not the quality of life.

Corporations in the United States were originally set up as public trusts, chartered by the states under strict regulations to fulfill public purposes. But this brief breakthrough into a different metaphor for economic activity was not to last. In time, corporations outgrew their masters and were granted the legal fiction of "persons."[8]

The Fallacies of Economic Thought

As the nation-state system and the capitalist economy grew more powerful, the frames that shaped the way people interacted through them were reinforced by the intellectual disciplines that emerged to explain and justify them. In order to legitimize themselves, the social sciences felt compelled to mimic Enlightenment physics. But the metaphor for economic activity—that of the lone rational actor competing in a self-regulating market—bore little resemblance to any previous or then current economic system, and it bears even less resemblance to the complex and volatile international corporate web that exists today. Ironically, such competitive self-interest was thought to lead to a state of equilibrium and the greatest good for the greatest number. In reality, the history of capitalism has been one of successive booms and busts—"creative destruction" (Schumpeter 1975), during which millions of people are violently displaced, communities are destroyed, and the land ravaged. In such a formulation, acting against one's self-interest—that is, in support of the common good or the land's sustainability—becomes irrational (Lakoff 2008, 209).

Like the nation-state, the metaphor for the market is that of a container, within which land, labor, capital, and technology (called "factors of production") are turned into products, services, and profits. Elaborate mathematical models were devised by economists to explain and predict how markets operate, but they have been built on metaphors and assumptions that have little correspondence to the messy, complex, contingent, and unpredictable quality of real life. As Nadeau has pointed out, the mathematical models developed by the creators of the neoclassical

Figure 1
Beginnings of Capitalism

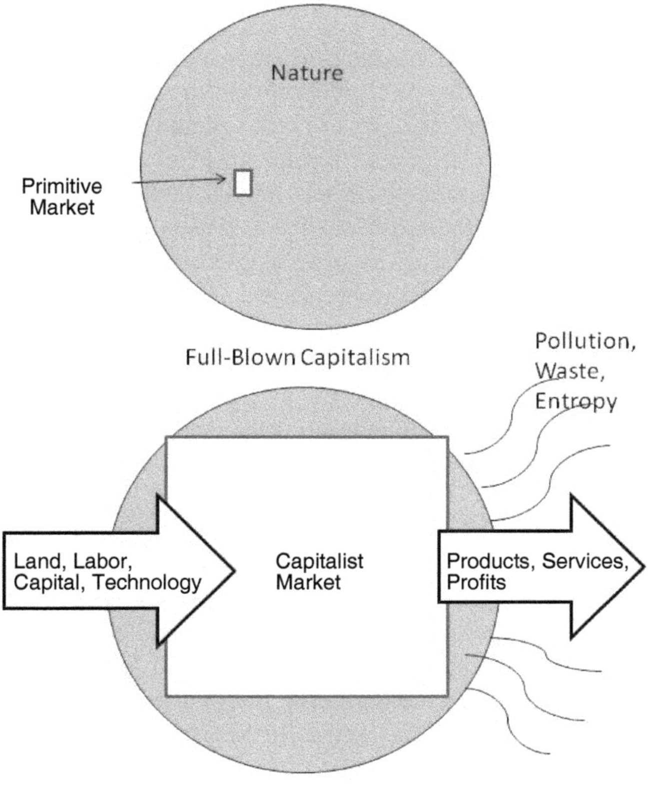

economic theory that serves as the basis for coordinating the global market system were based on a now obsolete theory that nineteenth century physicists used to account for the phenomena of heat, light, and electricity. The strategy they used, says Nadeau, "was as simple as it was absurd—they substituted economic variables for physical ones" (Nadeau 2008).

In their theory, the costs of environmental damage are treated as "externalities" and excluded from the pricing system that operates inside the box. Moreover, the model assumes that nature's resources are largely inexhaustible, and that those that are not can be replaced by others, or by technologies that minimize the use of the exhaustible resources or that rely on other resources. In effect, there are no biophysical limits to the growth of market systems (Nadeau 2008). Growth has come to be the defining dogma of markets—the secular religion of advanced and advancing industrial societies (Speth 2008). Yet nature's resources, under the law of growth, are not inexhaustible, as Figure one attempts to show.

The Fallacies of International Relations Theory

Like neoclassical economic theory, International Relations (IR) theory purports to be "scientific." It explains and predicts relations between nations. In IR theory, nation-states are both containers—and autonomous persons. They act to keep out "foreign threats" and compete against other nations to maximize their self-interest, which is taken to mean the public's interest, though in reality it is usually the interests of the nation's most powerful classes, even in ostensible democracies. In the most common model that game theorists devised to guide decision making in foreign policy, it is assumed that every nation acts to maximize its control over territory, natural resources, markets, economic growth as measured in GDP, and weaponry. The narrative results in a win/lose scenario. The assumption that every other nation is out for the same thing has led to tremendous destruction through economic competition and war.

The end of World War II suggested the possibility that a new paradigm of governance at the international level might emerge from the ruins of over a decade of depression and war. The establishment of the United Nations was an important breakthrough that enabled the international community to establish a body of international law that has greatly expanded our conception of human rights. But from its very inception, the U.N. was prevented from going further by the outmoded contours of Enlightenment epistemology. Nation-states set up the U.N., and governments get to negotiate with each other, with only a few governments having veto power over decision making in the Security Council. Moreover, the enshrinement of national sovereignty prevents the international community from having the power to effectively enforce its decisions; and the publics of these nation-states have no standing in U.N. deliberations. Moreover, indigenous peoples, because they are marginalized within their respective states, are entirely silenced in the councils of climate change deliberation when it is they (as Lankard, for example, demonstrates) who have the most to teach us about how to live in harmony with the earth. In addition, the realm of production, consumption, distribution, investment, and credit—the economic realm—has remained in a separate sphere, entirely outside U.N. jurisdiction.[9] Free to operate according to market principles, transnational corporations have created a new non-territorial transnational "space," a new international division of labor (Ruggie 2004, 7; Sassen 2008), and a powerful *private* international governance regime to regulate international trade, investment, and credit (Ruggie 2004, 7-8).[10] Such an arrangement has resulted in the exponentially increasing commodification and destruction of the natural "capital" upon which humanity and all of life depend (Ruggie 2004, 6).

Toward an Environmental Ethic of Global Governance

Human beings must order their collective lives around a radically altered worldview or paradigm based, not on Enlightenment science and philosophy, but on the findings of modern environmental science, quantum physics, and neuroscience. In effect, what we need is a new "Natural Law" to govern our interactions, not only with each other, but with the earth—a law that draws its principles and values from the structure of the earth system itself. While the task of operationalizing this new worldview has been taken up by others in this volume, I have suggested a schematic framework [Figures 2a-2d] that might serve as a conceptual guide. It contrasts the operative images/metaphors of a deep ecological framework or paradigm with the images/metaphors of the old Enlightenment paradigm. When this deep ecological perception becomes part of our daily awareness, a radically new system of ethics emerges (Capra 1996, 11; Korten 2006).

Getting from Here to There

While the need for a massive cultural paradigm shift is apparent, the question remains, how do we get from here to there? We may not have

Figure 2a

Enlightenment Worldview		Deep Ecological Worldview	
CHARACHTERISTICS OF NATURE			
Objective metaphors		Objective metaphors	
dead **inert** **unconscious** **impersonal** **predictable** **regular** **linear** **valueless** **useless** **waste** **wild**	Nature is made up of hard substances or particles that are inert or "dead" and are moved by external rather than inherent forces.	**alive** **conscious** **structure** **organization** **process** **a network of networks** **a community of communities** **interrelated** **cooperative** **self-creating** **self-regulating** **self-transforming** **contingent** **circular** **relational** **empathetic**	Nature is composed of energy that is infinitely varied in its behavior and properties. The world is "alive."
	Nature operates according to laws that are orderly, predictable, and self-perpetuating. Humans can use these laws to manipulate and transform nature for the benefit of the human species.		The laws of Newtonian physics are applicable only to an intermediate range of phenomena. They cannot predict events in the realm of very large or very small phenomena.
	Levels of predictability about the future can be achieved through understanding nature's law.		It isn't possible to make a complete prediction about the future. The very nature of reality is that we can only know one aspect of it at a time with any precision.
	Nature can be separated into its constituent parts in order to be studied and manuipulated. The whole is equivalent to the sum of the parts.		Nature is made up of a complex system of systems. The whole is more than the sum of the parts. What happens in one part of the system affects every other part.
	Whatever can be isolated, measured and quantified is the only valid means of discovering the "truth."		Quantification and isolation of phenomena is a limited means of comprehending reality. We must understand how systems operate.
	Nature has no intrinsic value, but only as it can be manipulated for the enhancement of human prosperity and happiness. "Land that is left wholly to nature is called . . . waste." (Locke)		Nature has an intrinsic value apart from its usefulness in increasing human prosperity and easing human pain. The human emploitation of nature creates waste.

Figure 2b

CHARACTERISTICS OF HOMO SAPIENS			
Objective metaphors		Objective metaphors	
autonomous **free** **rational** **purposeful** **above nature** **self-interested** **competitive** **conqueror of nature**	Human beings are basically isolated, selfish, competitive individuals but endowed with reason. Through the application and manipulation of nature's laws, humans can free themselves from the chaotic "forces of nature."	**social/ relational** **determined/ free** **rational/ emotional** **empathetic** **part of nature** **the universe** **becoming** **conscious** **of itself** **embodied** **embedded** **member** **diverse**	Human beings are the most complex life form, but exist only in interdependence with other lifeforms. They cannot achieve complete autonomy from nature.
	The purpose of human life is to achieve individual happiness (usually defined as wealth, or property), because such accumulation is the measure of one's intelligence, talent, etc.		The purpose of human life is to care for and live in harmony with eacher other and the earth.

Figure 2c

CHARACTERISTICS OF ECONOMY			
Objective metaphors		Objective metaphors	
self-regulating **rational** **autonomous** **quantifiable** **efficient** **closed box** **profit-oriented** **immortal** **value-neutral**	This private ownership of property (value extracted from nature) is a God-given right, and people have a duty to generate wealth because in doing so thye "increase the common stock of mankind" (Locke). The more one individually makes, the more society to collectively benefits.	**democratic** **self-sustaining** **self-creating** **reciprocal** **embedded** **living** **diverse** **interdependent** **a network**	Locally rooted, self-organizing communities of production consumption and exchange. Common ownership and shared democratic management of the global commons so that all benefit.
	Economics occupies a separate "sphere" and operates according to the "law of supply and demand" (Smith).		Production and exchange is embedded in the social structures and values of self-organized communities.
	The "invisible hand" of the market assures the proper balance between production and consumption and a fair distribution of the earth's wealth.		People do not take from the earth more than they can replenish. The health of future generations is embedded in economic principles.
	All phenomena that pass through the market are valued in terms of their quantity: cost per unit; price per pound; dollar per hour; wages per week.		Economic cost-accounting includes such non-quantifiable values are sustainability, aesthetics, durability, the costs of pollution, and the social costs of productive processes.
	Progress = quantity. A nation's wealth is defined as GNP or the sum total of all the goods and services produced in any given period.		Progress = abundance of clean air, water, healthy food, long-term sustainability, the fulfillment of human potential, time for family, community, recreation and creativity, the enjoyment of beauty.

Figure 2d

CHARACTERISTICS OF THE STATE/GOVERNANCE SYSTEM			
Objective metaphors		Objective metaphors	
box **container** **sovereign** **exclusive** **rational** **interest-maximizer** **autonomous** **controlling** **legitimate enforcer**	Humans freely "contract" with each other to form governments whose sole function is to protect the God-given or "natural" right of individuals to develop and exploit nature and thus to acquire property or wealth. (Locke)	**democratic** **interdependent** **a coomunity of communities** **a network of networks** **living** **self-regulating** **self-creating** **reciprocal** **embedded** **diverse**	Democratic governance is responsible for regulating the use of nature so that all may benefit.
	States are self-interested, power-seeking rational acotrs, who seek to maximize their security and chances of survival.		Governance regimes are voluntary, democratic, flexible and exist to promote the common welfare and to promote the common welfare and to sustain life on earth.
	States are defined by territory, language and culture.		Governance regimes vary in size and purpose and are dependent on/embedded in variety of biomes.
	In the interests of survival, states have a legitimate monopoly over the use of force.		Collective security rests in the reunciation of force and the provision of life-sustaining goods and services.
	Conflict is the norm in interstate relations.		Cooperation and reciprocity are the norms in intergovernance relations.

time for such a world transforming process to occur. We may have to settle for transitional strategies while working on longer term transformation. Or perhaps a precipitous global catastrophe will force us to make the more radical changes that are needed. While I do not have an answer to this conundrum, I would like to bring George Lakoff back into the picture to suggest that neuroscience may provide some clues. In addition to bringing the unconscious metaphors and narratives that have led us to the present crisis to consciousness in our political practice, there are other things we can learn for our political practice by the way the mind works. He lists the following:

Dealing with Traumatic Ideas

Neuroscientists find that ideas are instantiated in our brains, cementing synaptic changes under trauma and repetition. Lakoff cites as illustration the Bush administration's framing of the September 11 terrorist attacks on the United States as the "war on terror." The frame was then repeated endlessly so that it became physically instantiated in the brains of most Americans who came to use the phrase reflexively (Lakoff 2008, 128). Lakoff suggests that we need to find ways to inhibit ideas that have been developed as a result of traumatic experiences and to provide alternative modes of thought about the event—*to engage in reflective cognition about it*. Thinking in terms of how the nation-state has socialized us to think of those who threaten us as "enemies," peace and human rights activists might begin to reverse this habit of mind by repeating that those we perceive as enemies are part of the world's great web of natural diversity from which we might have something to learn, and that our own nation's past behavior might have something to do with the hostile or threatening reaction we are now experiencing.

The series of natural disasters that have been occurring with greater intensity around the world are finally awakening us to the dangers of the climate crisis. But we face the danger that the traumatic experiences through which we are made aware of the danger of climate collapse could inhibit us from seeing creative ways of responding. The fossil fuel industry, abetted by its political supporters, is now using these traumas to frame climate change as such a foregone catastrophe that the only thing we can do is learn to adapt. This then becomes an excuse not to adopt the more radical changes in governance and economy that are needed to slow the emission of greenhouse gases. They are also using the trauma created by crisis to enact policies that would otherwise be publicly unacceptable (Klein 2008).[11] Instead, policy makers could be using the findings of the IPCC to frame climate crisis as an opportunity to develop a more positive way of life, one which would bring the benefits of more sustained and cohesive communities; new employment opportunities; healthier air, water, and food; and more time for family, recreation, and creativity.

Framing Reality

"Sometimes," Lakoff says, "you have to construct a conceptual frame and a name when there are none, so that an important truth can be seen" (Lakoff 2008, 115). Vandana Shiva, for example, coined the phrase "biopiracy" to describe the privatization and commodification of the earth's genetic and intellectual inheritance by transnational corporations (TNCs) (Shiva 1977). [12] Following Lakoff, we could start referring to the practices of TNCs that destroy living ecosystems and cultures as "ecocide," and countries that engage in war and preparations for war must be accused of committing ecocide. It is indicative of our present dilemma that when war is discussed, the environmental consequences are almost never mentioned.[13] Through successively naming the truth, the human community may be brought to codify the crime of ecocide in international law, just as it has codified the crime of genocide. Indeed, since humans are part of the earth system, the crime of genocide would be a subset of the crime of ecocide.

A society that wants to make peace should also eschew the myriad ways in which campaigns to get rid of something we do not like, such as hunger, poverty, or disease are framed as a "war against" rather than as a campaign for shared prosperity or health. Indeed, too many cultures today are saturated with metaphors of war.

Another example of reframing would be to break the conventional conflation between economic growth and national well-being. Numerous studies have shown that people in countries in which a large GDP is taken to be the measure of progress do not experience happiness (Green New Deal Group 2008, 32; Gutrich in this volume). In fact, growth in the quality of life has actually diminished. "Growth," therefore, should be identified with a society's ability to provide more clean air, water, healthy food, parks and other wilderness areas, time for family, community, recreation, and the arts.

Overcoming the Fear of Framing

Lakoff points out that those who see the problems we face and want to work toward alternative solutions are often afraid to change the conceptual frames through which the situation is understood and discussed. A major part of this reluctance, he argues, stems from reformers' adherence to the eighteenth century idea of that if you just give people the facts they will act on them. The result is that progressives fall into the trap of being caught in the other side's frames (Lakoff 2008, 125-132). This fear is found among many environmentalists who understand the gravity of the climate crisis we are facing but fall into the trap of accepting the frames of the fossil fuel, nuclear, and auto industries. For example, they tell us that radically reducing our dependence on fossil fuels will bring the global economy to its knees, that solar and other renewable energy sources are too expensive to be a major source of energy for the foreseeable future, and that coal can be made to be "clean." Policy makers in the developing world have been trapped by the frame that the only way to bring a country out of poverty is to follow the development model of the industrialized countries. Such narrow framing of the issues prevents policy makers from seeing the alternative scenarios that are available and mires us in the trap of having to make false choices between the lesser of two evils.

Confronting Stereotypes

Policy makers often base their arguments on the use of stereotypes: making a particular member of a conceptual category stand for the category itself. (Lakoff 2008, 159-162). The fossil fuel industry and its cronies in the Bush administration vilified members of the radical ecology movement who destroy the property of multinational corporations or burn fields of genetically modified crops as "terrorists." This fringe movement is then made to stand for the entire environmental community and attention is diverted from the serious issues the group is trying to bring attention to through its actions. A more subtle form of stereotyping is found in the fossil fuel industry's attempts to discredit the findings of the IPCC. In their arguments, "science" is made to stand for the eighteenth century Enlightenment's view of accurate predictability. Al Gore's attempts to alert the global community to the immanence of global warming, for example, were discredited as being "unscientific" because a number of his predictions could as easily be attributed to natural patterns. Gore was then made to stand in for the entire community of climate change scientists whose work was deemed "unscientific" because of the wide range of uncertainties their findings present. The only true "scientists," therefore, were those who held that climate change is a hoax.[14] Those who are concerned about climate change need to learn to decode such stereotypes and to explain to the public how they work.

Conclusion

The power of Enlightenment thinking is beginning to weaken around the world and the institutional structures that embedded its metaphors and narratives are being reconfigured. Social movements have become important players in U.N. negotiations. International movements of indigenous peoples and peasant farmers have gained voice for a different understanding of and relation to the land, a more participatory form of governance to that of the sovereign nation-state system,[15] a participatory form of economic production and exchange,[16] a different concept of "development," [17] alternative forms of communicative relations and education, and a different basis for claiming rights protections (Collins 2008, Couch 2001, Moreira 2008, Zibechi 2007).[18] These movements have been challenging nation-state leaders and TNCs in numerous venues across the world (Bandy & Smith 2005, Speth 2008). The Earth Charter, a declaration of fundamental principles for building a just, sustainable, and peaceful global society designed "as a common standard by which the conduct of all individuals, organizations, businesses, governments, and transnational institutions is to be guided and assessed" is perhaps the best expression of the new ethic or "natural law" that has come out of the civil society movements (Earth Charter International). The Charter has been endorsed by a growing number of groups across the world and it is hoped that its promulgation will eventually result in its official adoption by the international community.

Nowhere has the implementation of such visions been more concrete than in South America, backed in some places by constitutional protections. Ecuador's new constitution, for example, is the first in the world to provide nature legal rights.[19] (Boudin, Gonzalez & Rumbos 2006, Collins 2008, Couch 2001, Moreira 2008, Zibechi 2007, Wilpert 2007).[20] Cuba has learned to produce a self-sufficient, organic food system with almost no fossil fuel inputs (Green New Deal Group 2008, 31). It is also in South and Central America that a new vision of international relations has been born. The ALBA or Bolivarian Alternative to the Americas, a plan for regional integration spearheaded by Venezuela, whose membership now includes Bolivia, Cuba, Dominica, Honduras, Nicaragua, and Venezuela, calls for reciprocal trade relations between member states based on cooperation instead of competition, and solidarity and complementarity instead of comparative advantage, with the goal of fighting poverty and inequality (Alternativa Bolivariana para los Pueblos de Nuestra América). Even in the industrialized countries we can see breaks beginning to happen in the pattern: cross-state and cross-national compacts to reduce greenhouse gas emissions; urban consumers turning to farmers' markets, etc.

Despite these and many other positive openings, the stranglehold of the Enlightenment box on the imaginations of the world's major decision makers remains strong, and the twin challenges of terrorism and global financial collapse may be dampening hopes for moving more rapidly toward the deep ecology paradigm. The balance sheet for the environment does not look bright, given the time frame within which climate scientists tell us we must act. Nevertheless, it may be that the sociopolitical economic order that has dominated the last 500 years of Western history may be just as subject to abrupt change as the climate. The sudden collapse of the world financial system in the summer of 2008 demonstrates just how fragile the system is; and without hope, we will do nothing at all. It is imperative that we begin to transform the frames, tell new stories, and share models of success wherever they are found.

Notes

1. Black Elk, Oglala Sioux, quoted in McLuhan, 42.
2. Chief Joseph, quoted in McLuhan, 54.
3. John Locke, Chapter V, Sec. 32, http://www.constitution.org/jl/2ndtreat.htm (accessed Sept. 5, 2008).
4. Other papers by Dr. Hansen can be found at his website: http://www.columbia.edu/~jeh1.
5. The authors, it should be noted, are careful to say that energy is not determinative of human history, but it does play a critical role.
6. For example, John Foster Dulles, Eisenhower's Secretary of State, George Schulz, Nixon's Secretary of Treasury and President Reagan's Secretary of State, Casper Weinberger, Reagan's Defense Secretary, Dick Cheney, George W. Bush's Vice President, and Condoleeza Rice, George W. Bush's Secretary of State, not to mention the Bush family itself, all had ties to energy companies. These are only a few of the myriad people with ties to the energy industry who have served in high office in the United States.
7. The tendency to objectify an "other"—usually persons outside the referential clan, tribe or nation—has, of course, been present throughout much of recorded history. Contemporary evolutionary biologists and psychologists, however, argue that humans are also social creatures with the need to cooperate for survival, and indeed throughout human history there are as many examples of human reciprocity as there are of its opposite.

8. Corporations were granted personhood in the 1886 Supreme Court case, *Santa Clara County v. Southern Pacific*. In the United States, corporations have been granted more rights under the 14th Amendment to the Constitution—the post Civil War amendment that was intended to provide citizenship rights for African Americans—than those for whom the amendment was originally written.
9. There have been some steps taken to bridge this gap. For example, the Global Compact is a U.N.-initiated compact between the corporate sector, civil society, labor, and governments based on principles drawn from the Universal Declaration of Human Rights, the International Labor Organization's (ILO) Fundamental Principles on Rights at Work and the Rio Principles on Environment and Development. But thus far, the compact is only voluntary and, according to Ruggie, there is little chance that the corporate community will subject itself to binding global regulations any time soon (Ruggie 2004, 31-32).
10. Given the fact that governments appoint representatives to sit on the governing bodies of organizations like the World Trade Organization (WTO), the World Bank, and the International Monetary Fund (IMF), one might say that the governance system is semi-private. Ruggie, for example, argues that TNCs still depend on state-mediated law for licenses to operate and to protect their property rights (Ruggie 2004, 7-8). However, this has been so since the rise of the modern capitalist economy. Government licensing and regulation of TNCs as well as government involvement in these transnational economic regimes masks the fact that representatives of the public have no ability to thwart the will of these bodies should they rule against the "public's" interest. The 1994 General Agreement on Tariffs and Trade makes democratically enacted national, state, and local laws subordinate to World Trade Organization rulings handed down by unelected panels that meet in private with no due process; corporations as well as nations may initiate WTO proceedings, and there is no way to appeal the rulings. While the IMF has lost considerable moral and financial clout in recent years, it can still demand that sovereign governments strapped by external debt conform to its structural adjustment regime at the cost of losing its access to foreign credit and investment.
11. Naomi Klein has demonstrated how neoliberal economists and their political allies have used the shock of crisis to impose disastrous neoliberal economic policies on countries and populations that might otherwise not accept such draconian measures.
12. Chalmers Johnson's (2000) use of the phrase "blowback" to describe the terrorist attacks against the United States is another example of the reframing being suggested.
13. The U.S. military is the largest single user of oil in the world, see Sohbet Karbuz, "The U.S. military oil consumption," *Energy Bulletin*, February 26, 2006, http://www.energybulletin.net/node/13199 (accessed January 12, 2009). Yet it is never acknowledged that the military may be the greatest producer of greenhouse gases and other pollutants, not only because of its consumption of vast quantities of oil, but because of the destructive power of its bombs and other weaponry.
14. Invariably, such scientists were often found to be in the employ of the fossil fuel industry.
15. Indigenous movements in Ecuador, for example, succeeded in getting written into the new constitution the concept of "plurinationalism," which would recognize the country's ethnic and cultural diversity and enshrine rights to protect this diversity of communities and nations within one nation-state. In theory, it would allow different groups to define, within limits, the rights they feel are inherent to their survival as cohesive communities.
16. Descriptions of such visions and practices can be found in numerous places. For a start, I recommend the Parecon section of ZNet, an online forum for the exchange

of civil society practices and visions: http://www.zcommunications.org/znet/topics/parecon.

17. Brazilian peasant farmers and environmental movements coined the term, *socio-ambientalismo*, or socioenvironmentalism, which stresses the need for development strategies to address both the needs of the poor as well as environmental ills at the same time (Angus Wright, "Is a better world possible? The experience of the Brazilian environmental movement and the "construction of citizenship." In Diez and Dwivedi, 289).
18. Collective or solidarity rights—for example, the right to peace, to a clean environment, to development, and to the right to one's language and customs—have not yet been incorporated into U.N. treaties but they have found recognition in some U.N. declarations and in the African Charter on Human and People's Rights and the Indigenous and Tribal People's Convention (ILO 1989).
19. The chapter granting rights to nature in Ecuador's constitution can be found at: http://www.celdf.org/ (accessed January 13, 2009).
20. The government of Venezuela uses a significant part of the oil revenues that used to go into the pockets of the elite to prepare its people, however unevenly, with the education, health care, and technical skills that will be needed for a post-oil based economy; organic agriculture and producer cooperatives are given preference in government grants and subsidies. The Venezuelan constitution, adopted in 1999, declares that the hydrocarbons found within its borders belong to the people and can never be privately "owned."

References

Alternativa Bolivariana para los Pueblos de Nuestra América. (n.d.), retrieved on 5 November 2008 from <http://www.alternativabolivariana.org/>.

Bandy, Joe and Jackie Smith (2001) *Coalitions across borders: Transnational protest and the neoliberal order.* Lanham, MD: Rowman & Littlefield.

Berry, Thomas (1988) *The dream of the earth.* San Francisco: Sierra Club Books.

Boudin, Chesa, Gabriel Gonzalez and Wilmer Rumbos (2006) *The Venezuelan revolution: 100 questions—100 answers.* New York: Thunder's Mouth Press.

Capra, Fritjof (1996) *The web of life: a new scientific understanding of living systems.* New York: Anchor Books/Doubleday.

Collins, Jennifer N. (2008) "Ecuador at the crossroads: rafael correa and the struggle for a new Ecuador." *Global Dialogue* 10, No. 1 (Winter/Spring).

Corcoran, Peter Blaze, Mirian Vilela and Alide Roerink (eds.) (2005) *The Earth Charter in action: toward a sustainable world.* Amsterdam: KIT Publishers.

Couch, Jen (2001) "Imagining Zapatismo: the anti-globalization movement and the Zapatistas." *Communal/Plural: Journal of Transnational and Crosscultural Studies* 9,No. 2 (October): 243-260.

Coulter, Karen (2007) *The rule of property.* New York: The Apex Press.

Daly, Herman E. and John B. Cobb, Jr. (1994) *For the common good: Redirecting the economy toward community, the environment, and a sustainable future.* Boston: Beacon Press.

Debeir, Jean-Claude, Jean-Paul Deléage and Deniel Hémery (1991) *In the servitude of power: energy and civilization through the ages*, trans. John Barzman. London: Zed Books.

Descartes, Rene. *Discourse on the method of rightly conducting the reason, and seeking truth in the sciences,* Chapter 6. English translation of *Discours de la méthode*, Project Gutenberg Literature.org E-Text No. 59. Retrieved on 20 August 2008 from <http://www.literature.org/authors/descartes-rene/reason-discourse/chapter-06.html>.

Díez, Jordi and O.P. Dwivedi (2008) *Global environmental challenges: perspectives from the south.* Peterborough, Ontario: Broadview Press.

Earth Charter International Council (n.d.) *The Earth Charter: A declaration of fundamental principles for building a just, sustainable, and peaceful global society in the 21st century.* Retrieved on 12 January 2009 from <http://www.earthcharterinaction.org/2000/10/the_earth_charter.html>.

Gelbspan, Ross (2004) *Boiling point: How politicians, big oil and coal, journalists, and activists have fueled the climate crisis—and what we can do to avert disaster.* New York: Basic Books.

Green New Deal Group (2008) *A green New Deal: Joined-up policies to solve the triple crunch of the credit crisis, climate change and high oil prices.* New Economics Forum: retrieved on 23 September 2008 from <http://www.neweconomics.org/gen/z_sys_publicationdetail.aspx?pid=258>.

Hartmann, Thom (2002) *Unequal protection: The rise of corporate dominance and the theft of human rights.* Emmaus, PA: Rodale Press.

Hyde, Lewis (2007) *The gift: Creativity and the artist in the modern world.* New York: Vintage Books/Random House.

Intergovernmental Panel on Climate Change (2007) *Climate change 2007: Synthesis report,* 2007. IPCC: retrieved on 15 June 2008 from <http://www.ipcc.ch/ipccreports/ar4syr.htm>.

Johnson, Chalmers (2004) *The sorrows of empire: Militarism, secrecy, and the end of the republic.* New York: Metropolitan Books/Henry Holt & Company, Inc.

______. (2000) *Blowback: The costs and consequences of American empire.* New York: Henry Holt & Company, Inc.

Klare, Michael T. (2004) *Blood and oil: The dangers and consequences of America's growing dependency on imported petroleum.* New York: Owl Books/Henry Holt & Co.

Klein, Naomi (2008) *The shock doctrine.* New York: Henry Holt & Company, Inc.

Korten, David C. (2006) *The great turning: From empire to earth community.* San Francisco: Berrett-Koehler Publishers, Inc. and Bloomfield, CT: Kumarian Press, Inc.

Kuhn, Thomas S. (1996) *The structure of scientific revolutions.* Chicago: University of Chicago Press.

Lakoff, George (2008) *The political mind: Why you can't understand 21st-century american politics with an 18th-century brain.* New York: Viking/Penguin Group.

______. (2007) Reply to Steven Pinker in "Mind games: does language frame politics?" *Public Policy Research,* (March-May): 59-71.

______. and Mark Johnson (1980) *Metaphors we live by.* Chicago: University of Chicago Press.

Locke, John (1690) *Second treatise of civil government.* Retrieved on 5 September 2008 from <http://www.constitution.org/jl/2ndtreat.htm>.

Ruggie, John Gerard (2004) "Reconstituting the global public domain: Issues, actors and practices." Working Paper No. 6, John F. Kennedy School of Government, Harvard University. Retrieved on 11 September 2008 from <http://www.hks.harvard.edu/mrcbg/CSRI/publications/workingpaper_6_ruggie.pdf>.

Moreira, Maria Elena (2008) "Human rights in Ecuador's new constitution." Retrieved on 15 September 2008 from <http://www.humanrightsmoreira.com/dhnceingles.htm>.

Sassen, Saskia (2008) *Territory, authority rights: from medieval to global assemblages.* Princeton, NJ: Princeton University Press.

Schumpeter, Joseph A. (1975) *Capitalism, socialism and democracy.* New York: Harper. (Originally pub. 1942).

Shiva, Vandana (2005) *Earth democracy: Justice, sustainability and peace*. Boston: South End Press.

______. (1997) *Biopiracy: The plunder of nature and knowledge*. Boston: South End Press,

Smith, Adam (1991) *An inquiry into the nature and causes of the wealth of nations*. Amherst, NY: Prometheus Books. (Originally published 1776.)

Speth, James Gustave (2008) *The bridge at the end of the world: Capitalism, the environment, and crossing from crisis to sustainability*. New Haven: Yale University Press.

Tarnas, Richard (1991) *The passion of the western mind: Understanding the ideas that have shaped our world view*. New York: Ballantine Books.

Wilpert, Gregory (2007) *Changing Venezuela by taking power*. London: Verso.

Zibechi, Raúl (2007) "Social change and building the ties that bind." *NACLA Report*, 15 (May). Retrieved on 4 September 2008 from <http://nacla.org/node/1473>.

6

Climate Change Policy with a Renewed Environmental Ethic: An Ecological Economics Approach

John J. Gutrich

Mounting evidence indicates that humans are significantly influencing the global atmosphere, leading to global climate change. Nations strive to better themselves over time, and efficient allocation of resources driven by a well functioning market is often cited by economists as an effective means to achieve this goal. Yet markets can have failures that result in the misuse of resources (a less than optimally efficient use), leaving society worse off over time. Degradation of critical non-market ecosystem goods and services, such as a functional carbon cycle that creates a global climate considered relatively stable for the support of human life, can leave society in a worse condition in its efforts to achieve a better state. Climate change arises as a tragedy of the global atmospheric commons as individual nations and peoples ineffectively attempt to better themselves while degrading an essential global resource shared by all. Yet, using an ecological economics approach, this chapter highlights the possibility of moving towards sustainable development as we face global climate change. Safeguarding our global atmosphere will require concerted international effort and a renewed environmental ethic.

Global Climate Change as a Reality

In reality, lifting people out of poverty and creating a sustainable environment are not conflicting aims; these goals are actually mutually supportive in a multitude of positive ways.

– Roundtable on Poverty and Climate Change in the U.K. (2008)

Global climate change is happening and growing evidence indicates that humans are contributing significantly to global climate change by releasing extensive amounts of greenhouse gases into the atmosphere. The Intergovernmental Panel on Climate Change reports that "warming of the climate system is unequivocal, as is now evident from observations of increases in global average air and ocean temperatures, widespread melting of snow and ice and rising global average sea level, global GHG (greenhouse gas) emissions due to human activities have grown since pre-industrial times, with an increase of 70 percent between 1970 and 2004, and most of the observed increase in global average temperatures since the mid-twentieth century is very likely due to the observed increase in anthropogenic GHG concentrations" (IPCC 2007).

With the industrialization and economic growth of nations, greenhouse gas emissions have increased greatly over time as many nations have utilized energy intensive production to fuel growing economies. Affluent nations such as members of the G8 (Canada, France, Germany, Italy, Japan, Russia, the United Kingdom, and the United States) release relatively large amounts of greenhouse gases. In 2004, the United States alone produced about 22 percent of global carbon dioxide emissions from burning fossil fuels (EIA 2008). Recent research also indicates the magnitude of China's GHG emissions, estimating that China surpassed the United States as the largest global emitter of carbon dioxide in 2007 (Auffhammer & Carson 2008). As the World Resources Institute (2008) highlights, however, while China and the United States emit approximately the same amount of greenhouse gases, China's per capita emissions are only 20 percent of those of the United States.

Global climate change will adversely impact people and cultures across the world by affecting critical freshwater supplies, land use, and human settlements affected by sea-level rise, food security, and human health risks linked to heat stress and increased transmission of tropical diseases (Harper 2008, Bates et al. 2008). Increased scarcity and competition for critical natural resources essential to meeting basic human needs are likely to create conflict and threaten political stability and peaceful relations within and across nations (Fingar 2008). The G8 nations and key developing countries, including China and India, have now acknowledged global climate change as "one of the great global challenges of our time" and G8 nations have pledged to try to halve greenhouse gas emissions by 2050 (Abramowitz 2008). Yet, with growing understanding of the potential significant adverse impacts of global climate change, why do nations engaged in global trade and market economies continue to prolong implementation of GHG mitigation efforts to pursue development paths now, which can ultimately undercut the objective of achieving a better standard of living?

In organizing the international conference *Facing Climate Change with a Renewed Environmental Ethic*, the Toda Institute, with its global mission of promoting peace, directly addressed climate change as a "great global challenge" by calling together scholars from across the world to formulate ideas and actions for effectively addressing climate change and promoting peace. This chapter discusses the problem of global climate change from an economic perspective and highlights an ecological economic approach to climate change policy and action based on a renewed environmental ethic.

The Global Climate Commons and Market Failures

The global atmosphere is inextricably linked to the survival of the myriad species that have evolved over millennia on the planet. The effect of the atmosphere on global climate is critical for the existence and continued survival of humankind. One can view the atmosphere as a critical resource for society that serves as both an environmental source of essential chemicals and thermoregulation and as an environmental sink where society emits chemicals that circulate through biogeochemical and hydrologic cycles. So once aware of the global climate change problem and potential devastating damages, why do nations continue to allow the alteration of a functioning and supportive atmosphere to a state that is causing global climate change?

Global climate change represents a classic example of a tragedy of the commons. The global atmosphere is a non-exclusive public good and the assimilative capacity of the atmosphere to process GHG emissions is a rival good. There are significant costs of climate change due to altering the chemical composition of the global atmosphere beyond its assimilative capacity and the costs do not affect any one nation entirely. Yet, benefits of the burning of fossil fuel as an energy source can accrue specifically to nations. Thus, a nation-state would logically continue to burn fossil fuels as long as the benefits accrued were more than the cost of climate change to that particular nation alone. Under this logic, there would result a race to destruction as each nation attempts to become better off from burning fossil fuels that result in positive *net benefits* to a particular nation, but in the end alter the global atmosphere (common resource) by ignoring the cumulative costs of emitting greenhouse gases. Furthermore, nations that were incurring large external costs of global climate change without the benefits of burning fossil fuels would now have a perverse incentive to ramp up fossil fuel production without an agreement on the shared use of the global atmosphere in order to increase net benefits. For example, remote Pacific island nations that reap relatively few benefits from extensive global carbon emissions face large potential external social costs from GHG emissions and global climate change. Island nations such as Nauru and Kiribati face rising sea levels that literally threaten the existence of the ground they walk on and the freshwater they drink due to salinization of wells by rising waters.

Overexploitation of public goods and common pool resources is not a new issue for society, as evident in other global resource issues such as fisheries and rangelands. In theory, markets serve to allocate resources efficiently and optimize net social benefits given firm property rights that allow persons consuming a good or service to express demand with a willingness-to-pay and producers with a willingness-to-accept to supply the good or service. Market failures occur when social costs such as climate change are not internalized and thus the level of production (energy derived from high carbon-emitting sources) indicated by a market may not be optimal for society. Economists often propose privatization in the case of depletion of a common property resource as a means to internalize costs back to the person(s) causing damage, thereby forcing the decision-maker to incur the true costs of the action. However, creating distinct property lines for non-exclusive goods and services such as fisheries and the global atmosphere is difficult if not impossible and thus requires further options for effective management of the shared resource (Ostrom et al. 1999). So why have nations failed to acknowledge that markets do not account for the cost of climate change, and that an international collaborative effort is required to maintain a global atmosphere that supports humanity? This may require an overview of how many nations assess progress and the economic development of the nation through markets.

Economic Growth as an Indicator of "Progress"?

Nations inherently strive to increase well-being over time and increasing gross economic production through increased energy and resource consumption has been utilized as a dominant strategy to try to achieve improvements in societal well-being. As the World Bank (2003) indicates, global energy use traditionally has grown at the same rate as Gross Domestic Product (GDP). GDP is a measure of the annual market value of final goods and services purchased in a nation (plus all exports net of imports). Research within societies and comparative analyses initially indicated a strong relationship between the growth of energy production and increases in measures of economic growth (Cook 1971, Harper 2008). Sociologists examined the relationship between energy growth and measures of social well-being (such as health, education, and nutrition), confirming a positive relationship between energy growth and growth in indicators of social well-being (Mazur & Rosa 1974). However, further studies did not support this notion of a positive relationship between growth in energy intensity and social welfare, as for instance, a case study comparison between the United States and Sweden (Schipper and Lichtenberg 1976).

In recent years, many scholars and economists have increased criticism of utilizing Gross Domestic Product (GDP) as a proxy for societal well-being. GDP counts *all production*, including activities related to environmental pollution and crime, while ignoring income inequality (Cobb et al. 1995, Talberth et al. 2007). Clean up expenditures of major oil spills are counted as an increase in the GDP, as well as harvesting forests beyond sustainable yields. Natural resource depletion is not effectively accounted for and thus short term gains of liquidating natural resources beyond natural growth rates or assimilative capacities to process pollution are highlighted as progress with no consideration of the user costs and sustainability of the production. Ironically, economist Simon Kuznets, chief architect of the GDP, informed the U.S. Congress seventy-five years ago that the "welfare of a nation can scarcely be inferred from a measurement of national income" (Kuznets 1934, Talberth et al. 2007). Prominent economists have reiterated this point that "GDP is not a measure of welfare" (Nordhaus & Tobin 1972) and "GDP can be a hopelessly misleading index of human well-being" (Dasgupta 2005). Scholars have also worked on alternative measures such as the Index of Sustainable Welfare (ISEW, Daly 2005) and the Genuine Progress Indicator (GPI, Talberth et al. 2007), two examples of studies that take into account natural resource depletion and income inequality.

University of Cambridge economist Partha Dasgupta supports use of the indicator of *wealth* per capita with "*wealth*" defined as the value of an economy's entire productive base comprising human capital (such as knowledge, skills, and institutions), human-made capital (roads, bridges, and buildings), and natural capital (forests, fisheries, and the global atmosphere) (Arrow et al. 2004, Dasgupta 2005). Dasgupta has indicated that figures published by the World Bank for the depreciation of certain natural resources (oil, natural gas, minerals, the atmosphere as a sink for carbon dioxide, and forests as sources of timber) indicate that in sub-Saharan Africa both GDP per capita and wealth per capita have declined in the past three decades, while on the Indian subcontinent, even while GDP per capita has increased, wealth per capita has declined (Dasgupta 2005). Ecological economist Herman Daly (formerly at the World Bank) has argued that sustained economic growth in a finite system is not possible and nations are trying to achieve higher GDP with increased production while liquidating natural resources (i.e., natural assets) that are critical complements in production (Daly 1992a, 1992b, 2005). Daly supports sustainable economic "development," which is an increase of societal well-being over time with resource efficiency, versus unsustainable economic "growth," which is economic activity fueled by the expansion of more resource/energy throughout (Daly 1996). Sustainable development is achieved with a dynamic economy that increases the qualitative state of people (skills, knowledge, etc.) and humanmade capital (roads, bridges, buildings, etc.) while maintaining productive natural ecosystems (natural capital) to support the economy (Daly 1996).

In terms of energy use, Harper (2008) highlights that macro-level studies and historical data for middle income developing countries identified two main phases of development and energy use: rapid industrialization and consumption highly dependent on increased use of energy from fossil fuels to economic growth becoming less energy intensive with a shift towards service industries and increased energy efficiency. This logic supports the idea that countries that have already industrialized and have increased income can now continue to raise national incomes with energy efficiency and subsequently lower fossil fuel emissions per capita.

In certain cases, resource economists have observed an empirical relationship suggesting pollution rising with income up to a certain point and then falling after some threshold, forming an inverted U-shape relationship (Shafik & Bandyopadhyay 1992, Barbier 1997, Auffhammer et al. 2002). A hypothesis was purported that levels of environmental degradation follow an inverted U-shaped pattern with increasing national income—observed by economist Simon Kuznets for income inequality. This relationship was termed an Environmental Kuznets Curve (see Figure 1). Unfortunately, the notion of an Environmental Kuznets Curve (EKC) for pollutants tends to fuel the idea of "grow first and clean up later," and that all a nation needs in order to solve pollution problems is to increase income rather than focus attention on the need to address sustainable development and an effective environmental policy (Arrow et al. 1995, Auffhammer et al. 2002, World Bank 2003). Strong inverted U-shape relationships with income have been observed for local air quality (sulfur dioxide and particulates) (World Bank 1992, Grossman & Krueger 1995, Cole et al. 1997, World Bank 2003), but income turning points for CO2 emissions are high (Schmalensee, et al. 1998) and studies of per capita emissions of CO2 have observed a steady worsening as per capita income increases (Arrow et al. 1995; Holtz-Eakin & Selden 1995; Easterly 1999). Keohane and Olmstead (2007) indicate that the empirical evidence for an EKC is weak, but there is also a lack of evidence that environmental quality necessarily declines with economic development, leaving the option for society to develop and maintain valuable environmental resources.

Figure 1 compares the Environmental Kuznets Curve hypothesis with empirical data of CO_2 emissions and increasing income levels across countries worldwide in 2000. There is no indication of an inverted U-shaped Environmental Kuznets Curve for carbon emissions, as CO_2

Figure 1

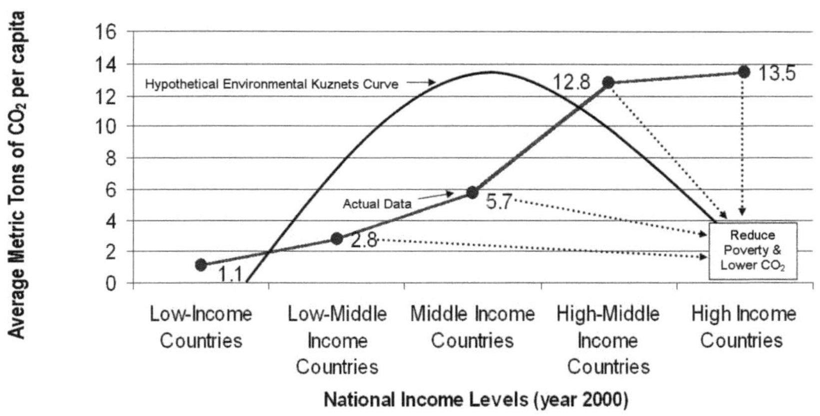

Note: Average CO2 emissions per capita (in metric tons) across national income levels in comparison to a hypothetical Environmental Kuznets Curve (EKC; after figures by Roberts and Parks 2007, p. 147). Dashed arrows (→) represent striving towards the objective of lowering CO2 emissions while reducing poverty and indicate 'leapfrogging' the EKC or lowering already high CO2 emissions. Information derived for the year 2000 for 192 nations with data from the Carbon Dioxide Information and Analysis Center (CDIAC, 2003) and the World Bank (see Roberts & Parks 2007, p. 147 and p. 284, note 42). Respective numbers of countries in each income level in the dataset include: low income (61), low-middle income (54), middle income (33), high-middle income (20), and high income countries (24; Roberts & Parks 2007).

emissions continuously grow with increasing income levels displaying high pollution levels in high income countries. The World Bank has reported similar trends in carbon dioxide emissions per capita with averages in 2004 of 0.9 metric tons in low-income countries, 4.0 metric tons in middle-income countries, and 13.2 metric tons in high-income countries (World Bank, 2008). Affluent countries are not displaying reduced carbon emissions, but rather are among the biggest polluters.

Interestingly, the United States under the Bush administration supported utilizing a "carbon intensity indicator (CO_2 emissions/GDP)" as a measure of the impact a country is having from carbon emissions, energy use, and economic growth that indeed displays an inverted U-shape relationship with rising income levels (Roberts & Parks 2007). Unfortunately, the measure is illogical because it directly trades carbon emissions for dollars of GDP growth without considering the benefits of the GDP growth versus the economic costs of carbon emissions. The only identifiable logic of the indicator is to support the notion that GDP growth provides greater benefits to society than the costs of carbon emissions and that nations simply need to grow GDP to lower the negative effect of carbon emissions. Unfortunately, to grow the GDP under current development approaches requires extensive energy and carbon emissions. The inverted U-shape relationship of the carbon intensity indicator is primarily an effect of the large denominator of GDP, and thus rich countries will have low carbon intensity as long as their economies are large and continue to grow rapidly. Therefore, while the proposed "carbon intensity indicator" declines with GDP growth, in reality the problem of global climate change becomes much *worse* with ever increasing *total* carbon emissions. Furthermore, the inadequacy of GDP as a welfare indicator highlights the perversity of policy based on a carbon intensity indicator.

Cleanup costs of environmental degradation are treated as increases in GDP, so the carbon intensity indicator would drop as long as a country was increasing its GDP by degrading its environment (i.e., with oil spills) and then using resources to clean up.

Nations often try to rationalize the strategy of continuing high levels of fossil fuel combustion (and subsequent GHG emissions) by coupling the logic of the ever-increasing economic growth model and the Environmental Kuznets Curve hypothesis—thereby reserving pollution abatement and cleanup for the future. In a single year, from 2006 to 2007, China added generating capacity that was equal to the whole of France's electricity grid and plans to build 500 coal-fired power plants in the next decade (Lim 2007). The United States has failed to join international agreements to curb greenhouse gas emissions in the short term and remains one of the top two emitters of CO_2 in the world alongside China. The United States has cited the exclusion of China from GHG cutbacks as rationale for lack of U.S. participation with international protocols and yet the U.S. has promoted an ever growing national production fueled by the highest per capita GHG emissions in the world (WRI 2008).

Major environmental problems arise with economic development based on an ever growing economy and the logic of an Environmental Kuznets Curve at the national and global level—as highlighted by global climate change. First, the EKC assumes that environmental pollution damage is reversible and that levels of degradation can be lowered or mitigated through future actions. In the case of the global atmosphere and climate change, economic development that causes extensive pollution at first, may not present the option to clean it up later if the degradation is irreversible (Solomon et al. 2009). The long residence time of carbon dioxide in the atmosphere (approx. 100 years) poses significant challenges and leaves limited options for future pollution abatement (Roberts & Parks 2007). Second, if one views the EKC only at the national level, specific nations may seem to be lowering environmental impact within the country, but in essence the environmental pollution is now released by other nations as the producers of products consumed through global trade. Again, ignoring cumulative impacts of nations will lead to development that fails to consider the true social costs of fossil fuel consumption and the effects of climate change.

Recent developments provide hopeful evidence that climate change policy may be changing for the top two largest emitters of carbon dioxide in the world. The World Resources Institute (2008) reported that China is now implementing an aggressive energy intensity target and a national renewable energy standard that highlights policies aimed at slowing greenhouse gas emissions growth. In the United States, the Obama administration has indicated support for regulating *total* carbon dioxide emissions and has asked the U.S. Environmental Protection Agency to review the policy of allowing states to regulate carbon dioxide emissions with more stringent auto standards than federal standards. Linking natural resource depletion, national security, and adverse impacts of global climate change, President Obama stated, "I want to be clear from the beginning of this administration that we have made our choice: America will not be held hostage to dwindling resources, hostile regimes and a warming planet" (Thomas et al. 2009).

Neoclassical Economics versus Ecological Economics

Most economists agree that well functioning markets serve to allocate resources efficiently, but the failure to properly reflect true social benefits and costs can lead to market failures that result in a less than optimal state for society. The benefit-cost approach in economics is based on premises of anthropocentrism (nature that is of value to humans), consumer sovereignty (the consumer's willingness-to-pay as indicative of utility through relative tradeoffs), individual valuations that are budget constrained, and social valuation estimated from the simple algebraic summation of individual variations (Randall 1987, Hussen 2004). However, neoclassical and ecological economists vary greatly on issues of distribution and the scale of the economy and its relation to the issue of effective policy approaches to global climate change (Daly 1992a, Tacconi 2000).

Neoclassical economists view human capital (such as knowledge, labor, skills, etc.), humanmade capital (manufactured products such as bridges, buildings, roads, etc.), and natural capital (natural resources) as fully substitutable, so that the economic growth of nations is not hindered by the scarcity of natural resources because there exists the possibility to fully substitute for natural capital with human and humanmade capital. As Brown and Panayotou (1992) highlight, resources are limited, but resource use is infinitely squeezable and the correlation between growth and environmental degradation may be a spurious one. Neoclassical economists argue that consumers and producers respond to changing relative incomes, prices, and constraints. If market signals are allowed to reach individuals and include all the social benefits and costs of individual actions, then response by individuals to problems will be economically efficient and best for society (Coase 1960, Hussen 2004, Harper 2008). Concerning global climate change, however, individuals and individual nations are not acting in consideration of true social costs and benefits, and this results in inefficiency and global degradation of the atmosphere.

Figure 2 depicts a global economy as a subset of nature consisting of human, manmade, and natural capital (after Goodland 1992, Daly 1992b). The world has experienced a dramatic increase in the human population and of manufactured goods over time while natural capital has been depleted to create manufactured goods and services. For most of history, the amount of humans and humanmade capital was relatively small in comparison to abundant natural capital (Daly 2005). Now, in the twenty-first century, the human population is six times larger than two centuries earlier and global gross production of goods and services exceeds $50 trillion and is projected to reach $140 trillion by 2050 (World Bank 2003, 2008). Global environmental problems are now evident,

Figure 2

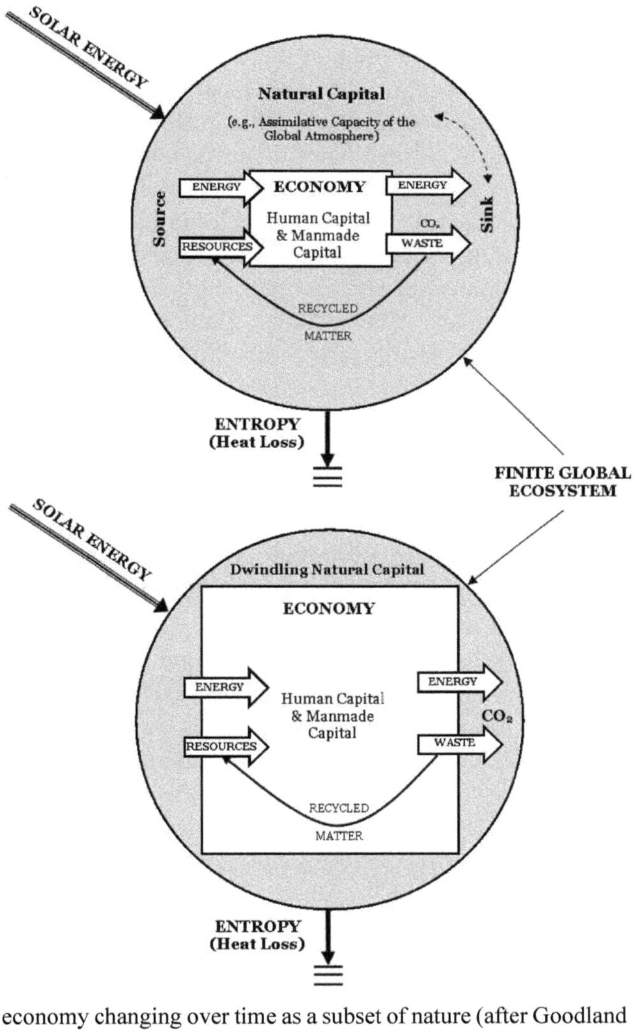

Note: The economy changing over time as a subset of nature (after Goodland 1992). An ecological economics approach to climate change requires maintaining critical natural capital (e.g. the assimilative capacity of the global atmosphere) to foster sustainable development.

including climate change, ozone depletion, desertification, and collapse of global fisheries. Neoclassical economists see little reason for concern with the depletion of natural capital over time because growing human and humanmade capital can substitute for nature to ensure increased well-being over time. Thus, in Figure 2, the economy (the box) could continue to grow indefinitely (outside of the circle) even with the depletion of nature (diminishing shaded area) because production could be fueled by humans and humanmade capital.

Ecological economists argue that there exists critical natural capital (see shaded area) that cannot be easily substituted and thus economic production will diminish over time if critical natural capital is depleted. Ecological economists believe natural capital and humanmade capital are complements in production and less natural capital will lower potential economic production and human well-being. Daly has effectively argued that one cannot have an economically productive fishery with only fishermen (human capital) and boats (humanmade capital), but no fish (natural capital; Daly 1992b, 2005). Ecological economists argue that we need to conserve the remaining finite critical natural capital and focus on sustainable economic development rather than sustained aggregate economic growth fueled in the short-term by liquidation of natural capital. In Figure 2, an ecological economic approach highlights the need to maintain the assimilative and thermoregulatory capacity of the global atmosphere as critical natural capital (shaded area) by changing behavior and economic production so that it no longer degrades this critical natural capital, but develops within natural constraints. In turn, environmental and developmental policy must acknowledge the constraints of nature, such as limited useful energy highlighted by the second law of thermodynamics, and shift towards efficient use of these limited resources in order to maintain and foster sustainable economic development. An ecological economic approach acknowledges that the loss of critical natural systems not captured in markets will, in the long run, hinder human development, and that international efforts should address policy that maintains these valuable natural assets.

Need for a Renewed Environmental Ethic and Effective Climate Change Policy

All arguments on the issue of global climate change, development, and natural resource use are fundamentally ethical. Utilizing benefit-cost analysis in economics to decide natural resource use and the effects on the global climate as a means to achieve development is inherently normative—based on society choosing "what ought to be." Philosopher Mark Sagoff (2000) highlights this point by arguing that economists apply benefit-cost analysis based on willingness-to-pay as a decision criterion for natural resources, but fail to use the same criterion to decide whether to use benefit-cost analysis to make such decisions. Resource economist Alan Randall (1987) has stated that "with respect to conservation and preservation issues, the arguments are ultimately ethical. Although a base of sound scientific knowledge is essential to identify the possibilities and predict the outcomes of alternative actions, decisions must be finally made on ethical grounds." Benefit-cost analysis in economics is an "unabashedly normative criterion" (i.e., subjective, based on human values) based on utilitarian ethical foundations (Randall 1987).

In the face of global climate change, it is apparent that a renewed environmental ethic is needed to maintain the critical functions of the atmosphere that support humanity. An ecological economics approach acknowledges that the global atmosphere is a form of critical natural capital for which there is virtually no substitute and which must be maintained in a natural state that supports humanity for the long term. Thus, an international effort and agreement is needed to establish such a renewed ethic and define a safe minimum standard (SMS; i.e., a minimum assimilative capacity of the global atmosphere for greenhouse gases) to maintain life supporting services. The concept of SMS was developed in order to address issues of species and biodiversity loss (Ciriacy-Wantrup 1952, Bishop 1978), but is applicable to climate change due to characteristics of very long time horizons, great uncertainty, and/or irreversibility (Randall 1987, Castle et al. 1995).

An ecological economic approach would set a global allowable level of greenhouse gas (GHG) emissions, but allow a market and tradable emissions permit system to optimize the net social benefits within the constraint of maintaining critical natural capital. Ecological economist Herman Daly argues such an approach *acknowledges the scale of our economy* in relation to remaining natural capital by setting a safe minimum standard in which the global economy can operate (Daly 1992a, 1992b). The mainstream economic growth model supported by high levels of fossil fuel consumption would need to shift towards a focus on energy efficiency to optimize social well-being to meet the requirements of the global agreement—and thus maintain a functioning global atmosphere.

Personal awareness of the social costs of greenhouse gas emissions and the contribution of GHGs to climate change can foster individual action, behavioral changes, and economic demand (of a type more closely reflecting optimal levels of consumption and production) that consider the full costs and benefits of fossil fuel consumption. However, without the full social costs of climate change reflected in market price, the markets will continue to mislead people and result in an overall state that is less than optimal for society. Thus, a renewed environmental ethic is needed with implementation of fewer GHG emissions at many scales of social organization, but implemented at the largest scale with an international agreement setting an upper limit by capping global GHG emissions.

This ecological economic approach based on a renewed environmental ethic calls for mitigation of the damage from global climate change with a global economy that operates within the natural capacity of the atmosphere to support all life. Nicholas Stern, former chief economist for the World Bank, clearly answered the question of whether the validity of this approach was worth it (i.e., cost effective) by reporting that in this century alone the damage from climate change could be *twenty times* the costs of solving it (Stern 2007). Investing just one percent of the global GDP over the next fifty years could stabilize GHG concentrations at about 25 percent above current levels, and yet failing to act could produce costs of $4 trillion by 2100 (Stern 1997, Harper 2008). The U.S. National Intelligence Assessment on the National Security Implications of Global Climate Change (Fingar 2008) indicated that climate change could affect: domestic stability in a number of key nation-states, the opening of new sea lanes and access to raw materials, and the global economy more broadly with significant geopolitical consequences. The U.S. National Intelligence Council report clearly states that "climate change will worsen existing problems—such as poverty, social tensions, environmental degradation, ineffectual leadership, and weak political institutions" (Fingar 2008).

The entire human population organized as a group of nations is the composite of billions of individuals and their actions. Even mainstream economists acknowledge the power of the individual as an independent consumer expressing her or his relative values for goods and services as a rational self-interested person in markets. Yet, ecological economists are willing to acknowledge that valuing and conserving certain critical natural services may require human decisions and the establishment of an environmental ethic outside of markets (and a willingness to pay) for the betterment of society. An ecological economic approach still respects the power of the individual in markets *along with* democratic agreement and implementation of an environmental ethic to protect critical natural capital.

Implementing Effective Climate Change Policy

Effective climate change policy based on a renewed environmental ethic will require implementation at multiple scales of social organization from local, regional, and national to international, but firstly and most importantly it will require an international agreement to cap global carbon dioxide emissions in order to eliminate the effects of "exporting" carbon dioxide emissions across nations. Calicott indicates that effectively addressing climate change will require a collective socio-cultural response in the form of policy, regulation, treaty, and law to offset the recalcitrance of fellow citizens that swamp personal efforts to lower carbon emissions (this volume). The same logic and concerns apply to groups of citizens at the regional and national levels. Nations and regions need assurances that development approaches will not be undercut or swamped by other non-cooperating nations releasing large amounts of greenhouse gases. Thus, international standards to limit global and national greenhouse gas emissions must be established to hinder personal actions of individuals, regions, and nations that will result in a tragedy of the atmospheric commons.

Once a limit on global greenhouse gas emissions is determined, the opportunity exists for strategic investments that serve to avoid the environmental damage of past inefficient development paths and to foster efforts to implement a renewed environmental ethic. An ecological economic approach with an international limit on greenhouse gas emissions would attempt to address carbon impacts and reduce poverty by "leapfrogging" the Environmental Kuznets Curve (or by greatly lowering greenhouse gas emissions if the EKC is not evident) and promote equitable economic development without extensive environmental degradation of our critical global atmosphere (see Figure 2).

The assimilative capacity of the atmosphere for carbon is at a threshold and the world can no longer afford economic development fueled by high carbon emissions. Yet, poverty and income inequality, inequitable historic emissions of carbon by now affluent countries, and high levels of vulnerability to climate change by poor regions that pollute the least, all highlight that climate policy based on a renewed ethic is also a matter of environmental justice. Westra warns us against expressing a diffuse concern for the immediate problem of climate change by speaking in a broad way of future generations, and emphasizes that we need to recognize that environmental justice concerns exist *now* as the first of future generations will come to be in our lifetimes (this volume). Balafrej highlights the vulnerability of the African continent and the global inequality of impacts, stating that "the regions that pollute the least are *not* those that are least vulnerable" (this volume). An effective international agreement must address the inequities of income across nations and propose carbon mitigation and adaptation efforts that attempt to lessen poverty and inequalities while lowering carbon emissions. Nanda reviews such potential for international environmental law to play a role in crafting appropriate mechanisms to support developing countries in their response to adverse impacts of climate change and clearly indicates that this is also a matter of "international human rights law" (this volume).

Goals of alleviating poverty and inequality and lowering greenhouse gas emissions are inextricably linked. Economist and activist Diez-Hochleitner indicates that poverty may be viewed as the world's worst "pollution" (Diez-Hochleitner and Ikeda 2008). Villamizar strongly argues that implementing a renewed environmental ethic for climate change involves a true conviction to eradicate poverty (this volume). Roberts and Parks (2007) in their work, *A Climate of Injustice*, argue that there exists the need to help nations upgrade their development pathways, diversify their exports to create stronger and more resilient economies, transition to lower carbon futures, and establish a *new* shared North-South worldview. A United Kingdom Roundtable Report on Poverty and Climate clearly indicated that a neo-classical economic approach will not be sufficient to accomplish such goals. He states, "[W]hat is clear is that tackling climate change simply through a price mechanism, without having a mechanism for transferring resources to the poor, will only worsen the already serious problem" (Johnson et al. 2008). An ecological economics approach to climate change acknowledges that carbon emissions and poverty levels can be lowered through effective policy, investment, and sustainable development that allow market mechanisms to determine efficient and cost-effective means to achieve international carbon caps and goals of poverty alleviation. The United Nations efforts towards promoting investments and local engagement in creating carbon neutral and climate change resilient territories, along with a collaborative role of non-governmental organizations (see Rogers, this volume), can foster development of low carbon municipalities, regions, and nations while trying to lessen the inequalities of historic inefficient development. Ultimately, lowering carbon emissions, poverty, and potential costs of climate change in turn lowers the risk for international conflict and promotes peace.

Conclusion

It only seems appropriate that the Toda Institute, founded by Daisaku Ikeda, philosopher, peace-builder, and educator, would coordinate an international scholarly effort and call for a renewed environmental ethic for climate change with Ikeda's notion that "[a] great inner revolution in just a single individual will help a change in the destiny of a nation and, further, will cause a change in the destiny of humankind" (Ikeda 1973). Climate change is a result of the cumulative impact of human-released greenhouse gases and alteration of individual behavior will ultimately be the prerequisite for mitigating global climate change. An international agreement based on sound science to limit GHG release and maintain a functioning global atmosphere would serve as the framework for an environmental ethic and as a call for individual action and ultimately international action. As Aldo Leopold (1949) indicated, an environmental ethic "reflects the existence of an ecological conscience ... and ... reflects a conviction of individual responsibility for the health of the land. Conservation is our effort to understand and preserve this capacity." An ecological economic approach to climate change supports market mechanisms and individual consumers to foster efficient means to achieve GHG limits, but acknowledges individuals can agree to an environmental ethic that highlights a conscience of critical natural capital by establishing a GHG emission standard that maintains a functioning global atmosphere for the betterment of humanity.

References

Abramowitz, M. (2008) "U.S. Joins G-8 Plan To Halve Emissions. 2050 Pledge Marks Shift on Issue for Bush." *Washington Post*, 9 July, p. A01.

Arrow, K., B. Bolin, R. Costanza, P. Dasgupta, C. Folke, C. S. Holling, B. O. Jansson, S. Levin, K. G. Maler, C. Perrings and D. Pimentel (1995) "Economic Growth, Carrying Capacity, and the Environment." *Science*, 268, pp. 520-521.

Arrow, K., P. Dasgupta, L. Goulder, G. Daily, P. Ehrlich, G. Heal, S. Levin, K. G. Maler, S. Schneider, D. Starrett and B. Walker (2004) "Are we consuming too much?" *Journal of Economic Perspectives*, *18*(3), pp. 147-172.

Auffhammer, M., R. T. Carson and T. Garin-Munoz (2002) "Forecasting China's carbon dioxide emissions: A provincial approach." *Second World Congress of Environmental and Resource Economists*, Monterey, CA.

Auffhammer, M. and R. T. Carson (2008) "Forecasting the path of China's CO2 emissions using province-level information." *Journal of Environmental Economics and Management*, *55*(3), pp. 229-247.

Barbier, E. (1997) "Introduction to the Environmental Kuznets Curve: Special Issue." *Environment and Development Economics*, 2, pp. 369-381.

Bates, B.C., Z.W. Kundzewicz, S. Wu and J.P. Palutikof (eds.) (2008) *Climate Change and Water*. Technical Paper of the Intergovernmental Panel on Climate Change, IPCC Secretariat, Geneva, pp. 210.

Bishop, R. C. (1978) "Endangered species and uncertainty: the economics of the safe minimum standard." *American Journal of Agricultural Economics*, 60, 10-18.
Brown, L. and T. Panayotou (1992) *Roundtable discussion: Is economic growth sustainable? Proceedings of the World Bank Annual Conference on Development Economics.* Washington: World Bank, pp. 353-362.
Carbon Dioxide Information and Analysis Center (CDIAC), Marland, G., T.A. Boden and R. J. Andres (2003) "Global, regional and fossil fuel CO2 emissions," in *Trends: A Compendium of Data on Global Change*. Oak Ridge, TN: Carbon Dioxide Information Analysis Center, Oak Ridge National Laboratory.
Castle, E. N., R. P. Berrens and S. Polasky (1995) "The Economics of Sustainability." *University Graduate Faculty of Economics. Lecture Series #8*. Corvallis, OR: Oregon State University.
Ciriacy-Wantrup, S. W. (1952) *Resource Conservation*. Berkeley: University of California Press.
Coase, R. (1960) "The Problem of Social Cost." *Journal of Law and Economics*, 3, 1-44.
Cobb, C., T. Halstead and J. Rowe (1995) "If the GDP is Up, Why is America Down? Why We Need New Measures of Progress, Why We Do Not Have Them, and How They Would Change the Social and Political Landscape." *Atlantic Monthly*, October.
Cole, M. A., A. J. Rayner and J. M. Bates (1997) "The Environmental Kuznets Curve: An Empirical Analysis." *Environment and Development Economics*, *2*(4), pp. 401-416.
Cook, E. (1971) "The flow of energy in an industrial society." *Scientific American*, 224(3), pp. 134-147.
Daly, H. E. (1992a) "Allocation, distribution, and scale: towards an economics that is efficient, just, and sustainable." *Ecological Economics*, *6*(3), pp. 185-193.
Daly, H. E. (1992b) "From Empty-World Economics to Full-World Economics: Recognizing an Historical Turning Point in Economic Development," in R. Goodland, H. E. Daly and S. El Serafy (eds.), *Population, Technology and Lifestyle: The Transition to Sustainability*. Washington: Island Press, pp. 170.
Daly, H. E. (1996) *Beyond Growth: The Economics of Sustainable Development*. Boston: Beacon Press.
Daly, H. E. (2005) "Economics in a full world." *Scientific American*, *293*(3), pp. 100-107.
Dasgupta, P. (2005) "A measured approach." *Scientific American*, *293*(3), 106.
Diez-Hochleitner, R. and D. Ikeda (2008) *A Dialogue Between East and West: Looking to a Human Revolution*. New York: I.B. Tauris and Co. Ltd.
Easterly, W. (1999) "Life during growth." *Journal of Economic Growth*, *4*(3), pp. 239-275.
Energy Information Administration (EIA) (2008) *Greenhouse Gases, Climate Change and Energy*. U.S. Department of Energy. May 2008: DOE/EIA-X012.
Fingar, T. (2008) *National Intelligence Assessment of the National Security Implications of Global Climate Change to 2030*. Statement for the Record of Dr. Thomas Fingar, Deputy Director of the National Intelligence for Analysis and Chairman of the National Intelligence Council. Before the House Permanent Select Committee on Intelligence and the House Select Committee on Energy Independence and Global Warming, June 25, 2008.
Goodland, R. (1992) "The Case the World Has Reached Limits," in R. Goodland, H. E. Daly and S. El Serafy (eds.), Population, *Technology and Lifestyle: The Transition to Sustainability*. Washington: Island Press, pp. 170.
Grossman, G. M. and A. B. Krueger (1995) "Economic Growth and the Environment." *Quarterly Journal of Economics*, *110*(2), pp. 353-377.

Harper, C. L. (2008) *Environment and Society: Human Perspectives on Environmental Issues*. 4th edn., New Jersey: Prentice-Hall.

Holtz-Eakin, D. and T. M. Selden (1995) "Stoking the Fires? CO2 Emissions and Economic Growth." *Journal of Public Economics*, 57, 85–101.

Hussen, A. (2004) *Principles of Environmental Economics: Economics, Ecology and Public Policy*. 2nd edn., Routledge, pp. 464.

Ikeda, D. (1973) *The Human Revolution*. 1st edn., Weatherhill.

Intergovernmental Panel on Climate Change (IPCC) (2007) *Climate Change 2007: Synthesis Report*. (Valencia, Spain, 12-17 November 2007).

Johnson, V., A. Simms and C. Cochrane (2008) *Tackling climate change, reducing poverty. The first report of the Roundtable on Climate Change and Poverty in the U.K.*. London: New Economics Foundation, pp. 40.

Keohane, N. O. and S. M. Olmstead (2007) *Markets and the Environment*. Washington: Island Press, pp. 274.

Kuznets, S. (1934) "National Income, 1929-1932." Senate Document No. 124, 73d Congress, 2d session, 1934.

Leopold, A. (1949) *A Sand County Almanac*. Oxford: Oxford University Press.

Lim, L. (2007) *China's Coal-Fueled Boom Has Costs*. National Public Radio. 2 May. Retrieved from <http://www.npr.org/templates/story/story.php?storyId=9947668>.

Mazur, A. and E. A. Rosa (1974) "Energy and lifestyle: Cross-national comparison of energy consumption and quality of life indicators." *Science*, 186, pp. 607-610.

Nordhaus, W. and J. Tobin (1972) "Is growth obsolete?" In *Economic Growth*, National Bureau of Economic Research Series No. 96E. New York: Columbia University Press.

Ostrom, E., J. Burger, C. B. Field, R. B. Norgaard and D. Policansky (1999) "Revisiting the Commons: Local Lessons, Global Challenges." *Science*, 284, pp. 278-282.

Randall, A. (1987) *Resource Economics: An Economic Approach to Natural Resource and Environmental Policy*. 2nd edn. New York: John Wiley and Sons.

Roberts, J. T. and B. C. Parks (2007) *A Climate of Injustice: Global Inequality, North-South Politics and Climate Policy*. Cambridge, MA: Massachusetts Institute of Technology.

Sagoff, M. (2000) "Environmental economics and the conflation of value and benefit." *Environmental science & technology*, *34*(8), pp. 1426-1432.

Schipper, L. and A. J. Lichtenberg (1976) "Efficient energy use and well-being: The Swedish example." *Science*, 194, pp. 1001-1013.

Schmalensee, R., T. M. Stoker and R. A. Judson (1998) "World Carbon Dioxide Emissions 1950-2050." *Review of Economics and Statistics*, *80*(1), 15-27.

Shafik, N. and S. Bandyopadhyay (1992) "Economic Growth and Environmental Quality: Time Series and Cross Country Evidence." *World Bank Policy Research Working Paper WPS 904*. Washington: The World Bank.

Solomon, S., G. K. Plattner, R. Knuttic and Pierre Friedlingstein (2009) "Irreversible climate change due to carbon dioxide emissions." *Proceedings of the National Academy of Sciences*, *106*(6), pp. 1704-1709.

Stern, N. (2007) *The Economics of Climate Change: The Stern Review*. Cambridge University Press, pp. 712.

Tacconi, L. (2000) *Biodiversity and Ecological Economics: Participation, Values and Resource Management*. London: Earthscan.

Talberth, J., C. Cobb and N. Slattery (2007) *The Genuine Progress Indicator 2006: A Tool for Sustainable Development*. Oakland, CA: Redefining Progress.

Thomas, K., H. J. Hebert, D. Cappiello and Erica Werner (2009) *Obama orders push to cleaner, more efficient cars*. Washington: Associated Press (AP).

World Bank (1992) *World Development Report 1992*. New York: Oxford University Press.

World Bank (2003) Sustainable *Development in a Dynamic World. Transforming Institutions,Growth and Quality of Life*. World Development Report 2003. New York: Oxford University Press.

World Bank. (2008) World *Development Indicators 2008*. Washington: World Bank Publications. 418 pp.

World Resources Institute (WRI). (2008) *Energy and Climate Policy Action in China*. Washington: World Resources Institute.

7

Two Global Crises, Ethics Renewal, and Governance Reform

Andrew Brennan and Y. S. Lo

Two global crises characterize the present day: first, the deepening problem of climate change, which is a marker of a range of other environmental problems faced by the planet; second, the more acute crisis triggered by the financial and economic meltdown of 2008 and its consequences. Both, we argue, are crises of prosperity occurring within an economic system in which positive feedback amplifies a pathology of consumption which takes the form of an addiction to intrinsically unsatisfying goods and pleasures. Unless affluenzic addiction to consumption is controlled, then invocations to individuals to limit their consumption and live sustainably are likely to fail. This chapter explores the possibility of finding systemic solutions to the problems of overconsumption and addiction, but also warns that these problems may be so grave as to lie beyond solution.

Two Global Crises: Climate Change and Financial Meltdown

Humanity is faced with two global crises of prosperity. Each one raises questions about the predominant ways of life in contemporary consumer societies, and the ways in which our values, motives, behaviors, and capacities are molded, if not determined, by the various systems of transaction in which our lives are set.

The first is the deepening problem of climate change—one which until recently was met with corporate and government denial, confusion, and disorientation. Responses to scientific warnings have been widely held by scientists to be inadequate to the problems faced in a world with rising sea levels, failing crops, water scarcity, intensifying storms, and the other damaging effects of a changing climate. Likewise, radical critics and a minority of journalists themselves have long been critical of the biased reporting of climate change issues.

However, public opinion and beliefs on the issue have changed during the last ten years. The idea of contracting industrial countries' carbon emissions, and aiming for convergence, if not equity, across the globe in per capita emissions, now appears to have achieved some credence in both the mainline media and mainstream social and political movements of the rich industrial countries. Moreover, talk about carbon taxes and caps has entered the standard vocabulary of politics, with both US presidential candidates in the 2008 election advocating at least a 60 percent reduction below 1990 levels by 2050 in the United States' own carbon emissions. While conservation groups and many scientists have criticized the Australian government for advocating a greenhouse gas reduction of not more than 80 percent over 2000 levels by 2050, at least discussions about climate change have now entered the mainstream public debate in Australia after years of being the focus of only academics, scientists, and other minority groups.

Talk is one thing, action another. What we want to draw attention to here are the reasons given for action—the motives. The press release for the official Australian government review of climate change, conducted by economist Ross Garnaut states, "without strong, effective and early action by all major economies, it is probable that Australians, over the 21st century and beyond, will experience disruption in their enjoyment of life and increasingly of their prosperity" (Garnaut *Media* 2008). Debate around the report centered on levels of atmospheric carbon expressed in parts per million by volume (ppmv), with particular focus on whether targets should be 450 rather than 550 ppmv. The report itself indicates that at an atmospheric carbon target of 450 ppmv, the Great Barrier Reef would be severely bleached and damaged, while at 550 ppmv "there would be a disappearance of the Reef as we know it" (Garnaut 2008, 127). As far as species loss is concerned, the report also indicates that "the 550 ppmv outcome would lead to a greater incidence of species extinction. Under the expected temperature outcome from the 550 ppmv scenario, 12 percent of species are predicted to be at risk of extinction. This percentage is reduced to almost 7 percent under the 450 scenario" (Garnaut 2008, 11).

Not surprisingly, critics of the Garnaut report have pointed out that the environmental cost of the 550 strategy is unacceptably high. Clive Hamilton, one of Australia's best-known radical economists and green advocates, points out that the Garnaut review seems to lack even a ...

convincing economic rationale. According to Hamilton, using Garnaut's own figures, the economic difference between aiming for 550 ppmv rather than 450 ppmv by 2050 is actually very small. Australia's GNP is set to double, Hamilton argues, by 2040. By adopting the 550 strategy, that doubling is deferred by two years to 2042. With a 450 ppmv target, the doubling will take just another six months beyond that. Over the whole century, the difference between the two strategies is economically negligible. Hamilton writes:

> Even to ask whether it's worth waiting an extra six months for our incomes to double in exchange for a much better chance of averting climate chaos is bizarre. To conclude, as Professor Garnaut has, that the trade-off is not acceptable to the Australian community is possible only if economic growth has become a fetish.
>
> Does anyone believe that Australians would be less happy if they had to wait an extra six months before they became twice as rich? The absurdity of the situation set out in the Garnaut report suggests our obsession with economic growth is so powerful that we are unwilling to contemplate sacrificing a tiny amount of consumption to sharply reduce the risk of irreversible damage to the Earth's climate (Hamilton 2008).

What provokes the ire of Hamilton, the Australian Conservation Foundation, and a host of other critics, is the tone of a report in which economic measures are used as the only key measure of worth. Among the terms central to Garnaut's analysis are "enjoyment of life" and "prosperity." The principal "costs" of climate change and global warming are taken to be economic ones, associated with "revenue," "tourism," and lost opportunities for "growth." The last few decades of environmental ethics have seen the language of *intrinsic value*, *moral standing*, and *rights* being deployed as a way of representing why natural things matter beyond human utility and consumption. While writers have argued that what is at stake in conserving them is of greater moment than human prosperity, Garnaut's report is silent on all these topics. It counts the "non-market values" associated with the environment as being worth separate listing and noting only. Such values are not conceived as providing any reasons or motives for government policy and action. In fact, the report has real difficulty in dealing with "non-market values": it is as if the author lacks the language in which to speak about or represent them properly.

There is also a second, apparently more acute, crisis of human prosperity, one immediately concerned with material goods, pleasure, and wealth. This is the global economic meltdown, beginning with the collapse of financial markets around the world in the second part of 2008.

It started with the failure of the sub-prime loans market, which then triggered a cascade of credit freezes, market shocks, takeovers, and bankruptcies that has not only led to the partial nationalization of previously private companies and banks, but also to further use of the language of catastrophe. Terms like "meltdown" and "Armageddon" have been used by journalists to capture the sense of panic associated with the massive downward movements in stock markets all over the world.

Government injections of liquidity into national financial systems all over the world have shown that even in an era of intense globalization, the nation-state is alive and well, and capable of taking significant economic, financial, and political action. This would hardly be worthy of comment were it not for a range of globalization theorists and academics who have been engaged in recent years in celebrating—rather prematurely—the demise of the nation-state (see, for an influential source of this view, Appadurai 1996). The financial crisis of 2008 has, among its many other consequences, led to the recognition that national governments have major roles to play not only in internal political and economic matters, but also in ways that affect the global economy. It has also been a stark reminder that there are many basic conditions for a healthy and flourishing industrial society. While theorists of many kinds have advocated the importance of a sustainable environment, social justice, open and democratic decision-making, and a whole range of other factors, the current financial crisis has drawn clear attention to the necessity for a reliable and responsible financial sector.

In a striking difference from the response to climate change, the funds voted to restore financial health and build up confidence in banks, and to finance houses and companies around the world, have been enormous and have been made available with impressive rapidity. In doing so, political leaders have themselves taken a stance on values obligations and social responsibilities, and some of them have also spoken in detail about such things. The Australian prime minister, Kevin Rudd, commented, "it is perhaps time now to admit that we did not learn the full lessons of the 'greed is good' ideology. And today we are still cleaning up the mess of the 21st Century children of Gordon Gekko." Declaring that "we've seen the triumph of greed over integrity," Rudd stated that the credit crisis revealed a lack of values among corporations and executives (ABC 2008). In a subsequent speech to the Press Club, the Prime Minister "denounced the comprehensive failure of extreme capitalism that caused the crisis and now turns to government to prevent systemic failure ... As a government, as a nation, we must respond to the twin evils which are of the root of this malaise: greed and fear" (Grattan and Medew 2008).

Systems of Positive Feedback

We believe that both the ecological and the financial crises are products of the same system of positive feedback loops. Locked inside these loops is a peculiarly single-minded kind of actor, one whose basic hedonistic features are exploited and perverted so as to feed the continual operation and growth of the system. These features are, of course, the central features of human nature, ones which are fed and exaggerated by the system under which many people now live. Globalization has effectively strengthened this already robust structure of positive feedback loops by rapidly expanding its territory of persuasion. Common labels for such a system are "consumerism" and "capitalism." Both labels miss a crucial aspect of the system, one that is best captured in the popular label *affluenza* (De Graaf, Wann and Naylor 2005) and its popular definition as an illness characterized by addiction, anxiety, depression, and *ennui* (James 2007).

The notion that we are infected by the system of commodities within which we live our lives, measure our status, and seek ever more elusive satisfactions, can be traced back to critical postmodern theorists such as Stephen Pfohl and the notion of the "parasite café" (Pfohl 1990). The term "affluenza" usefully captures the involuntary nature of both our exposure to, and damage by, the toxic structures that infect us. While the labels "capitalism" and "consumerism" capture the material aspect of the disease (particularly the gigantic scale of production and consumption of material goods, and the hugely accelerated imbalance in the accumulation of material wealth), the word "affluenza" points to the fact that the disease is suffered by the affluent in their addictive pursuit of capital and consumption, at the cost of destroying welfare, the environment, and individual health.

Affluenza is a product of systems that are rich in available resources, hence able to thrive and multiply on the addictions cultivated in consumers. At the same time, these addictions are focused by the system on a range of pleasures and satisfactions that are—in comparison to the capacities of normal human beings—relatively narrow. The result is an atrophying of higher sensitivities, a loss of the richness of lived experience, and its replacement by repetitive satisfactions based on material goods. Alongside this are symptoms of addiction—typically a failure to recognize the scale of the problems we face, optimism about our ability to get out of deeper and deeper problems, borrowing from the future, and an unrealistic and exaggerated sense of future profits and satisfactions.

All bubbles burst in the end, and so—we maintain—will the system that drives and promotes the frenzy of affluenza. The system is inherently self-destructive, and that means that if we do not find a means of controlling it, we will find ourselves the losers. Containing the environmental damage, the damage to our sense of community and dignity, and the other harms wrought by our present state of disease and dysfunction, will not be easy. But that, we argue, is the challenge for environmental ethics in an age of changing climate and financial crisis. The questions posed by the two crises just described are similar in uncomfortable ways. Just as no-one knows for sure what will revive a failing financial system, so no-one knows for sure whether the world can survive an ecological meltdown—a literal one, given the loss of glaciers and icepacks under the effects of global warming.

The difference in response to the two crises shows that at both global and national levels there is an asymmetry in thought, sentiments, and action about the two challenges. The financial collapses led to coordinated and urgent political action to try to save heavily indebted financial systems, even in the face of criticism from voters concerned about giving financial help to those who had been the agents of the collapse. Headlines complained about rewarding bad behavior and the spotlight turned on the huge salary payments and share allocations received by chief executives. Despite these doubts, the G7 finance ministers came together to endorse huge spending and showed their leadership by putting real money into collapsing systems.

By contrast, very little by way of real resources, or real amelioration has been pumped into the declining and much more seriously stressed global environment. This asymmetry requires some explanation. Many scientists, conservationists, and philosophers found it baffling that it was so hard to get species loss, water scarcity, pollution, climate changes, and other environmental crises onto the public agenda in the first place. They have wondered why there has been such a successful degree of evasion, denial, and procrastination about these things. Can it be because we can see our money suddenly lose value, while we do not in the same way see the slow disappearance of species or experience a sudden reduction in environmental resources? In the case of individual people, unfitness and failing health creep up in cumulative ways that avoid triggering alarms. Bankruptcy, or massive financial losses, can—like a heart attack—happen all of a sudden and trigger just the panic and fear to which Kevin Rudd makes reference.

While there is truth in this way of looking at things, a deeper account takes us to the point that money matters because consumption matters. As Zygmunt Bauman points out, we are now raised as consumers: consumption is our fundamental role or function in the political and economic world. Childhood, Bauman argues, is no longer a protected zone, an exemplar of a kind of human innocence, but instead a cog crafted to fit a huge, demanding scheme of consumption, marketing, and brand positioning. Think of how children now dress, make up and compare brands, and outdo each other in trendy attire. They present to us not as the emerging and unformed persons they will one day become, but instead appear as strangely diminutive adults. In place of the human qualities of childhood, Bauman claims, there has been a crude reprocessing aimed at producing a class of consumers trained in shopping dependence (Bauman 2005, 110 – 114). We are so well trained that we are unaware of our blindness to such remarkable features of the age.

Our claim is that the two meltdowns—the ecological and the financial—are the result of a system (or group of systems) within which we find ourselves trapped. Those systems are inadequately labeled as "consumerism," the "market economy," the "new capitalism," or by any other of the common labels. It is a system that corrodes and atrophies fundamental human abilities and sensitivities—those required for appreciation and value. Worse, the systems within which we find our place as consumers can actually reinforce the corrosion and decline of the very things that are of greatest value in human life.

The Atrophy of the Human

Here is how the spiral of reinforcing atrophy works. Let us begin from the idea that we are all consumers now, that for us, nothing is more important than consuming. The ideal "citizen" of the affluenzic society is someone for whom buying and acquiring material goods and marketable services is an all-consuming passion. Of course, as Hannah Arendt repeatedly emphasized, the pursuit of private goods and the discussion of how best to obtain them is not a proper concern of the public realm in a healthy society, hence our use of quotes around the word "citizen" in the last sentence (see Arendt 1958, 1968). While we agree with Arendt's analysis, our concern in this paper is to look at a further problem with the expansion of what she calls "the private realm," and its domination of contemporary life. When consuming is all-important, then things are ranked in terms of the private preferences of individuals seeking satisfaction. One loop in the system of reinforcing atrophy is now formed by the fact that individuals work in order to get the money through which their material demands can be satisfied. To get more and more goods and services, the individual has to make more money, and hence spend more time working. The imperative to consume thus gives rise to reduced time for other activities, including, paradoxically, reduced time for consumption itself. The reduction in time available to individuals, as we shall see, in turn reinforces the drive to consume more.

Of course, if consumption were an ordinary desire, focused on ordinary things, it would be unlikely to have the paradoxical effect of driving us to devote more time to work, when what we rationally want is more time for enjoying the other things that we desire. However consumption is not an ordinary desire, but a pathology: it is addictive, in just the way described in popular books about affluenza (see James 2007), and also in the way pointed out years ago by eco-psychologists (see Glendinning 1995 for an account that focuses on the consumption of, and dependence on, technology). Its pleasures are not intrinsically satisfying. Consumer goods, as such, are always updateable, upgradeable, upscalable. To update, upscale, and upgrade is always possible—so long as you have the money to do so. It is the fate of any consumer product to become unsatisfying and therefore need replacement. Consumer goods are not meant to last or provide long-lasting satisfaction. On the contrary, the ideal consumer product is something that people initially lust after, and then quickly become dissatisfied or bored with, ensuring that they will soon seek a replacement. Actually, most consumer goods are not intrinsically satisfying, nor are they meant to be. They are designed not to be, for that is how the system of consumption and production keeps running.

It is consumers' dissatisfactions with what they currently have, and their desire for, and belief in, better prospective products that maintain the flow of the market. It follows that successful product marketing must not only create desires and beliefs but must also create dissatisfaction by building its seeds into the very consumer products themselves. Accordingly, we have an everlasting cycle that runs from discontent through to desire, hope, belief, and then back to discontent. For many people, the cycle of yo-yoing emotions from the low ends of disappointment and boredom to the high ends of excitement, hopefulness, and optimism is a psychological equivalent to a physical addiction resulting from drug abuse. In all forms of addiction, the highs are short lived, and when they give way to the lows, the addict is compelled to seek the highs again to ease the pain, the void and emptiness inside. Affluenza is a form of addiction. The highs it offers and the lows it generates coordinate and feed into each other, so that the addict is locked into the cycle.

Once consumption is recognized as an addiction, this makes sense of the fact that we are quick to deny it (denial is a key feature of addiction). We also underplay its role in our lives, and falsely claim to be in charge of it, when in fact the reverse is true. Consumerism, seen in this way, explains two further apparent addictions: our addiction to work, and our addiction to credit.

Consumerism is an apparently safe, legal, and easy form of pleasure-seeking. However, once the consumer pattern of life is established, the offerings of the economic system themselves start to replace other forms of pleasure. Indeed, there is evidence that regular forms of pleasure—the joys of food, love, companionship, fresh air, country strolls, singing, dancing, and a host of other inexpensive joys—are increasingly less valued than the immediate forms of gratification available in the market place. The destructive effects of addiction to drugs, alcohol, food, and cigarettes are well known, and recent studies have suggested that in some populations, virtual relationships and sex toys are preferred to intimacy with other human beings (Sparrow 2008). While sex, food, and drink raise quite complex issues surrounding addiction and its causes, we conjecture that some element of these addictions reflects the spread of a wider addiction in consumer societies. Richard Sennett has coined the phrase "corrosion of character" to apply to what we have identified as the *atrophying* of values and ethics within organizations (Sennett 1998). The atrophying we have in mind, however, is widespread and occurs throughout the societies we inhabit, not just inside organizations. It involves a withering of sensitivity to everyday sources of pleasure that are characteristically human. As sensitivity withers, we become less and less able to appreciate the very joys that are central to a fully human life.

Indeed, the better the consumer, the narrower the kinds of desires that person should have. Here is the real price of the consumer society. Only the kinds of desires that can be satisfied by marketed goods and services are encouraged as proper for cultivation by consumers. Once other, non-market desires and sensitivities wither away, then sources of enjoyment become increasingly limited. However, unlike bodily changes due to aging and disease—ones that are often irreversible, ones that we have to cope with as best we may—the changes induced by consumerism are potentially reversible and able to be resisted. Without such resistance, though, and in a world where material goods are a prime source of value, the other sources of value become increasingly reduced and increasing wealth is not accompanied by additional happiness (see chapter four of Frey & Stutzer 2002). To an extent, then, our diagnosis supports E.F. Schumacher's gloomy view that "the result of the lopsided development of the last three hundred years is that Western man has become rich in means and poor in ends" (Schumacher 1977, 58).

Future Abuse

Compared with the "joys" of addiction, freedom and autonomy become less highly valued in societies where the consumer character is cultivated. Usefulness and durability are less important in the consumer society than the trendy, the latest model, and any other feature that is linked to the social approval that is craved by the addictive personality. The result is unneeded goods that are tossed out not long after being bought. We claim that such a culture encourages the junking of moral values like integrity, honesty, and decency—all the things that make human life so worthwhile. False and misleading advertising, overblown claims, and public relations hype all become accepted parts of our experience in advanced capitalist societies. Once truth, honesty, and sincerity are viewed with cynicism—a cynicism encouraged by every false promise in advertising—then these virtues become undermined, resulting in a further corrosion of both our ethics and our very humanity. Once intelligence, reason, and cleverness are detached from truth, honesty, and sincerity, disaster is not far away.

It should come as no surprise that financial institutions engage in irresponsible lending to people who have no likelihood of repaying their debts, and that it is the smartest guys in the room who drove Enron into a scandalous collapse (McLean & Elkind 2003).

Here, then is a two-level loop. Not only do consumer desires to feed our addictions lead us paradoxically into longer working hours, but they also rob us of the time to reflect on the loss of our sensitivities and abilities, and on the narrowing of our desires and values. There is an interesting link between consumption, stress, depression, and time starvation. The paradox of working more to get more not only drives the atrophying of sensitivities and corrosion of character, but also pushes us into high stress modes of life. What to do in the face of such inevitable sources of stress? One easy option is to use consumption itself as a salve for stress. There is evidence that stress and depression lead people to bad eating habits—high fat, salty, and sweet foods bring some quick solace to those who are unhappy and short of time. There is thus a link to bad health and the spread of lifestyle diseases typical of our age, which can be traced not just to the cheapness and ready availability of junk food, but to the more general role of food in helping consumers cope with anxiety, depression, and stress. Advising people to be more discerning, more reflective, and more careful in what they eat is likely to be an ineffective strategy for those locked into an addictive consumer system.

One common suggestion is that people should leave the affluenzic system of consumption—or at least reduce their participation in it—once they have enough to satisfy all reasonable desires. Such self-imposed constraining of material desires is called "downshifting." Self-help books on the topic advocate self-control. But the addictive desires inculcated by affluenzic societies are irrational and competitive, giving rise to status anxiety and other pathologies. Only a few people, relatively speaking, try to break free from the system, with optimistic estimates placing the ratio of downshifters in the 30 to 50 age group at around 20 to 25 percent in Australia (Hamilton & Dennis 2005). However, in the affluenzic culture, those who adopt an anti-consumer style are taking risks for themselves, their families, and relatives. Peer pressure and social expectations will often mean that individuals who would like to be less involved with the system of getting and spending will be careful about doing so, in case their choices inadvertently damage the lives of those near and dear to them.

The system of interlocking feedback loops we have described can simultaneously explain the two meltdowns—ecological and financial. The opposite of atrophy is hypertrophy, the excessive growth in the economy and the overconsumption of material goods that leads to environmental destruction. In the final analysis, the entire project of contemporary industrial and economic developments is one big scheme of borrowing with no security. We borrow from the planet's future reserves (the ecological goods and services that would be of value to humans and members of other species yet unborn) in order to satisfy our present desires for, and addictions to, consumption and profit making. We justify our huge borrowings on the credit that we think we have—a faith in human resourcefulness and the technologies that we hope we are going to invent for paying back all our debts. Furthermore, our borrowings from future generations are security-free, meaning that if we are unable to pay back the principal (let alone the interest) and if the natural environment goes bankrupt as a result, then too bad—there is no immediate loss to us. In short, we are simultaneously the borrower, the creditor, and the broker, all of whom will be long gone by the time the ecological meltdown occurs and the contemporary industrial bubble bursts. Exactly the same kind of unsustainable borrowing and lending in finance sectors all over the world has already produced very similar kinds of consequences: abuse of credit, hugely inflated prospective returns, reckless underestimation of investment risk, and finally the bankruptcy of giants that have been previously thought to be unsinkable.

Psychologically, the two crises are strikingly similar: the sacrifice of the future in pursuit of short-term prospective gains, which will themselves fail to satisfy us, so that we make further borrowings against the future to pursue yet further illusory gains, and so on. Put concisely, the system of dysfunctionality that is being explored here is a system of *future abuse*, both ecologically and economically. Note that there is a certain cleverness to the way the future eating is carried out. We take the ecological resources now that will be lost to future generations of humans, and to future generations of plants and animals. And people at the rich end of town speculate not with their own money, but with the mortgages, pension funds, and savings of those at the poorer end of town. In each case, our addictions involve us in gambling beyond our means, and gambling away wealth that belongs to others. By the way, gambling itself is another highly addictive form of activity. When different forms of addiction are pulled together and causally intertwined as they are in the system of affluenza, they grow on, feed off, and justify each other. When individuals in the system experience the lows of one addiction, they can quickly fix it by resorting to the highs of another addiction. There are no limits to the availability of the many forms of highs—except that every one of them is transient, intrinsically unsatisfying, and invariably brings the individual back to the lows of pain, emptiness, and frustration.

This model of interlocking systems that leads to the two different forms of future abuse is highly oversimplified and we do not claim to have done more than touch on the outlines of something extremely complex. We hope that even the present sketch reveals much of interest. Not only is there a lack of diversity among the services and products offered to people, but this is accompanied by a simplification of human sensitivities, desires, values, and even thoughts. Such simplification is reflected in other losses too, such as a narrowing of language, and the trivializing or emptying of substantial and useful notions such as "truth," "taste," "dignity," and "respect." Such a narrowing of human experience is at odds with the self-conceptions of those living in the industrial societies, and of many ethnographers, who use measures of cultural sophistication that regularly locate the researcher, and the inhabitant of modern industrial society, as more informed, flexible, articulate, artistically-sensitive—more sophisticated—than those who live in poorer countries, or in traditional societies. Our analysis, by contrast, queries this, and leaves it as an open question whether the impoverished or infantilized self of the consumer society should be placed lower on the scale of sophistication than someone in a traditional society who may, for example, enter into complex networks of diverse relationships with many other different people, and whose capacity for pleasure, enjoyment, and satisfaction has not been channeled into material satisfaction linked to profit based on unsecured credit.

Our analysis also places us in some gloomy company. In effect, we are arguing that in place of the intrinsic value of human dignity, the individual in the consumer society is seen in more instrumental terms as a means to achieve the goals of profit. In place of "man the measure," human beings are increasingly regarded as no more than wheels in a larger mechanism, not as instruments of biology spreading their genes (as the sociobiologists have argued), but as instruments of the markets seeking to make profits. We are not far, then, from the pessimism of John Gray when he tells us that "the modern myth is that science enables humanity to take charge of its destiny; but 'humanity' is itself a myth, a dusty remnant of religious faith" (Gray 2003, 4).

Beyond Solution?

Our analysis may seem to raise a problem that is beyond solution. Our model explains why appeals to rationality, self-control, care for biodiversity, the triple bottom line, and the other solutions usually given for the environmental ills of the age are all likely to fail. Our vision of the present global system and its national mirrorings is of a beast with a tendency to addictions, and to form human beings who themselves are prone to ever greater addictions. Marketing and branding are among the activities now devoted to capturing and enthralling the addiction-prone consumer and persuading him or her to purchase what the marketing agency wants people to buy.

If there is to be hope for the future, then we need to find some ways of building limits into a system that threatens to run beyond control. Herbivores are normally controlled in nature by their predators long before they eat themselves out of resources. But the artificial system of affluenza that we have unwittingly invented has no controls built into it. Think of it like this: the better the system is at producing consumer-junkies, addicted to consumption and the work that allows them to consume, then the better it is at making profits and strengthening as well as expanding itself, and then the more stubbornly addicted consumer-junkies it is going to produce, and so on and so forth. Given the facts of ecology, this ever expanding affluenzic system cannot run forever. While governments have assumed the responsibility for stabilizing and maintaining world markets by active political and monetary interventions, even more so—we suggest—should they take responsibility to control addictive overconsumption when such behaviors of their citizens threaten the stability of societies and the survival of the human species as a whole. We have time here to suggest just two interconnected lines of intervention: other lines are possible, and much needed.

As a general rule, the longer it takes for people to master something, the more difficult it is for them to unlearn it. By the same logic, the more time and effort that people have to spend in order to acquire the taste for a certain kind of pleasure, the longer it will take for them to get bored with the experience. That is why people become dissatisfied with consumer goods so quickly. For they are the exact opposite. Like bright treats for children, consumer goods are immediately attractive, easy to enjoy, and dangerously addictive, but quickly become dull, and harmful when consumed excessively.

This being so, the most obviously direct way to counter affluenza is to ensure that an important proportion of people in the society are equipped with counter-consumer capacities. In the exact opposite fashion to the consumer psyche, it is essential that the counter-consumer has the abilities to derive long-lasting pleasure and satisfaction from things and experiences that are economically inexpensive, but the acquisition of the taste for which requires time and cultivation. For reasons given above, people with such sensitivities and capacities will be naturally drawn to non-consumer goods and prefer the enjoyment of them to consumption.

One way of inducing the envisioned change in social demographics would be for governments to nurture new generations of citizens with counter-consumer dispositions and capacities, perhaps through national education systems and programs. Individuals coming through such systems of new non-consumer values will develop refined sensibilities and strong dispositions to appreciate and enjoy free or low-cost non-consumer goods and experiences, such as friendship, pursuit of knowledge for its own sake, country walks, games that do not involve high technology, the meeting of minds, visiting galleries, understanding of human affairs, solving logical puzzles, observing animal behaviors, singing alone, singing in groups, studying, creating art and music, dancing, self-reflection, and gentle conversation with like minds. In short, we need a good proportion of people to be equipped with the capacity to derive long-lasting satisfaction from non-consumer goods which have low environmental and economic costs.

There are strong prudential reasons for governments, and even corporations, to assist the production and nurturing of the counter-consumers as described above. And there are ethical reasons for doing so as well. The prudential reason is that unless the affluenzic system finds ways of limiting its own expansion, then the system will become self-destructive, and whatever good it has brought to the stakeholders will be lost. If nation-states and multinational corporations care for their own future existence, then they have a profound interest in ensuring that the affluenzic system in which they have been thriving does not end up destroying itself and everyone in it. The paradoxes of the affluenzic and addictive societies can—according to this line of thought—be countered by a move that itself seems paradoxical, a move governments and corporations may well be forced to take sooner rather than later in order to preserve the systems within which we live.

Living with Dignity

The ethical reason for building up resistance to consumerism is one that can be traced back to Aristotle, one that suggests that there are possibilities in the human character for a certain nobility and dignity that is diminished if we aspire to being nothing more than addicts. The thought appears many times in intellectual history. It has a famous expression in Pico della Mirandola's oration on the dignity of man, where he depicts human beings as poised between beasts on the one hand, and angels on the other, with the freedom to choose in which direction to go (Pico 1486). We can, says Pico, fall into the merely animal mode of life, one of immediate gratification. Incidentally, a critical reading of Pico would note that he attributes to the animals the very features of immature or undeveloped humans. If we are right in claiming that consumerism fosters the infantilized society, his remarks about animals can be re-read as remarks about our childishly addicted selves. What he says thus applies to the lures, temptations, and addictions about which we have been writing.

On the other hand, Pico insists, we can aspire to something that we may never fully attain, but toward which we can aim—the divine pursuit of wisdom, justice, and understanding, the pursuit that brings us nearer to heaven, that transcends the immediate. Such talk is not completely lost nowadays. There are already people who are successful in business and whose lives are not dedicated simply to working, getting, and spending, but who see their lives as affording them entry to a new kind of nobility or aristocracy—of letters, of aesthetics, of all the fulfilling pursuits that humans are capable of. For those who have the strength of character for it, the systems in which we operate can be rewarding—a reward that, regrettably, is based on the addiction and infantilizing of others.

Such aspirations inevitably raise the question of limits—the question whether there are limits to what humans can do and how far they are able to realize their dreams. As Pico and the other renaissance writers pointed out, human creativity and human goodness are both limited, not only by human nature, but also by the nature of the world itself. Much of modern discourse has been dominated, as John Gray and others have noted, by a belief in limitless progress and unbounded freedom (Gray 2003, 2007). These naively progressivist notions resonate well with the delusions of addiction that we have already outlined. By contrast, our own project involves recognizing the importance of limits—limits which define our own humanity and which will protect social and ecological formations from the damage that arises from the pursuit of limitless growth.

The ethical grounds for building limits into present systems of profit-making are based on hopes for human dignity and autonomy that are not grounded in delusions of infinite progress, nor founded in attempts to manipulate individuals into becoming anti-consumers, anti-spenders, and anti-debtors. Instead, we are looking for structural means within the present system for increasing the prevalence of forms of human nobility and dignity and for maintaining notions of respect for others. This is ostensibly what education already aims at—though radical critics of education bewail the fact that much of what goes on in secondary schools and universities is geared to producing administrators, bureaucrats, accountants, and managers and not to the development of human character. Existing educational institutions, and the maintenance of genuinely liberal education within these institutions, may still have the capacity to produce people with critical minds and high expectations—expectations that are not satisfied in the showroom, shopping mall, or department store. We have suggested ways of doing this in the previous section, and have mentioned easy and obvious structural means by which the current system could be changed so as to improve the lot of those who are at present victims of the atrophying effects of affluenza. They would aim to encourage the study and pursuit of philosophy, literature, politics, art, biology, pure science, various kinds of craftsmanship, and other activities that are not directly linked to the ideal of ever increasing consumption. Simple though they are, such changes would require the assistance of enlightened members of the business community and strong direction and regulation by the government. Consider the requirement that all corporations provide thirty percent of their profits to sponsor, encourage, and develop just such pursuits within the society as whole. It is hard to imagine any government now being brave enough to suggest such a thing—but this level of investment in non-consumption may just be what is required to save the consumer society from its own self-consumption. The support of non-consumption may be just what is needed to preserve a future of modest consumption.

One piece of mixed news for the environment emerges from the financial meltdown. Policies that would have been regarded as unthinkable, as too radical, as unpalatable to the national and worldwide business economy—such policies were introduced in a flash, thereby drastically changing the nature of banking in many parts of the world. With partial government ownership of major banks now achieved so quickly, we face a liberating moment, a time in which we can think of policies to regulate and control the destructive effects of affluenza in ways that would previously be discounted as utopian or extremist. This may seem like good news, but the message is actually mixed. It would be encouraging to think that human beings could take necessary evasive action before the obviousness of the environmental crises we face is brought home to us by a series of calamities. But the financial meltdown is also a sobering precedent for unheeded warnings, for slow response to signs of crisis in the system, and for putting off action until enormous damage has been done. The culture of affluenza, as we have already argued, is essentially a culture of denial, hence one that puts off and delays sensible preventive measures.

Governments often provide financial help to first home buyers, and to those starting up in business: in response to the financial meltdown, the Australian government increased its first-home grant, giving $21,000 to first-time home buyers. By contrast, such bonuses are seldom paid to those who want to ease out of the workforce, who opt for part-time work, who want to live gently on the planet, and who want to bring up their families with wider intellectual and experiential horizons than are provided for in the affluenza culture. In the spirit of thinking about strategies that may seem dangerously utopian, consider this further one: suppose governments started to underwrite the risk on the part of those people who want to reduce their impact on the planet, and who therefore want to change in radical ways their present involvement in the circle of increasing working hours, increasing stress and anxiety, and increasing consumption. For many people, this solution invokes the spectre of a radical kind of socialism—one that provides supportive social structures so that the families of those who reduce their engagement in the consumer society are still given the chance of a good education and the ability to have access to health, welfare, and other services. Such critics can perhaps be reminded that until recently the suggestion of partial government ownership of some of Europe's biggest banks would likewise have seemed too radically socialist to consider. From the point of view of saving some of the present system, while preserving the planet's ecosystems, we may have little choice but to investigate taking routes like these.

Conclusion

In the history of Europe, the sponsorship of talented musicians, scientists, medical researchers, writers, philosophers, and poets was often the hallmark of a certain greatness of mind—one that paid enormous cultural and social dividends. One recent book suggests that the absence of an Enlightenment in eighteenth century England was due to the lack of elite patronage—in contrast to the situation in both Scotland and France (Andrew 2006). Patronage itself was of central importance in developing independent thought, since it freed the great writers of the day—figures like Adam Smith, Jean-Jacques Rousseau, and Denis Diderot—from the economic need to write for a wide audience of paying readers. The Enlightenment age provides a startling contrast from the contemporary scene in countries like the United Kingdom and Australia. Public institutions and governments have not taken on the role played by the eighteenth century aristocrats, but instead there is a pressure for the arts, philosophy, and letters to pay their way, and encouragement is given for universities and industry to link up in joint research projects of national economic benefit. At a time of bold initiatives designed to maintain confidence in financial institutions, and maintain the sustainability of global economic systems, it is perhaps time to start considering equally bold moves to liberate people from the addictions of affluenza. The local and global environments on which our lives depend, whose operations we still do not understand, whose wonders are still capable of filling us with awe—these are worthy of urgent protection, and of brave and bold political action. This is a challenge not only for national governments and international organizations, but also for companies of all sizes. In the eighteenth century, it was the aristocracy who sponsored the arts and sciences, thereby laying the foundations for an age of human intellectual and technical development of an unprecedented scale. The question we now face is whether the new corporate aristocrats—those whose importance and power is based in massive corporate wealth—will have the same vision and inspiration as their eighteenth century counterparts, people whose influence and status was founded on inheritance or military prowess. If global climate change were to provoke a new ecological enlightenment, then there may be some hope for the future.

References

ABC (2008) "World failed to learn from Gordon Gekko: Rudd," ABC News, 6 October 2008. Retrieved on 12 October 2008 from <http://www.abc.net.au/news/stories/2008/10/06/2382689.htm>.

Andrew, Edward G. (2006) *Patrons of Enlightenment*. Toronto: University of Toronto Press.

Appadurai, Arjun (1996) *Modernity at Large: Cultural Dimensions of Globalization*. Minneapolis: University of Minnesota Press.

Arendt, Hannah (1958) *The Human Condition*. Chicago: University of Chicago Press.

Arendt, Hannah (1968) *Between Past and Future*. New York: Viking Press.

Bauman, Zygmunt (2005) *Liquid Life*. London: Polity.

De Graaf, John, Wann, David, and Thomas Naylor (2005) Affluenza: The All-Consuming Epidemic, 2nd edition. San Francisco: Berrett-Koehler.

Frey, Bruno, and Alois Stutzer (2002) *Happiness and Economics*. Princeton: Princeton University Press.

Garnaut Media (2008) *Climate Change Review*, Media Release 30 September 2008. Retrieved on 8 October 2008 from <http://www.garnautreport.org.au/reports/Media%20release%20-%20Garnaut%20Review%20Final%20Report%20-%2030%20September%202008.pdf>.

Garnaut, Ross (2008) *The Garnaut Climate Change Review.* Melbourne: Cambridge University Press.

Glendinning, Chellis (1995) "Technology, Trauma and the Wild," in T. Roszak, M. Gomes and A. Kanner (eds.), *Ecopsychology: Restoring the Earth, Healing the Mind.* New York: Sierra Club Books.

Gray, John (2003) *Al Qaeda and What it Means to be Modern.* London: Faber and Faber.

Gray, John (2007) *Black Mass*. London: Allen Lane.

Hamilton, Clive (2008) "What is the Future Worth?" September 2008. Retrieved on 8 October 2008 from <http://www.clivehamilton.net.au/cms/media/documents/articles/garnauts_tiny_costs.pdf>.

Hamilton, Clive and Richard Deniss (2005) *Affluenza.* London: Allen & Unwin.

James, Oliver (2007) *Affluenza*. London: Vermilion.

McLean, Bethany and Peter Elkind (2003) *Smartest Guys in the Room: The Amazing Rise and Scandalous Fall of Enron*. New York: Penguin.
Pico della Mirandola, Giovanni (1486) *On the Dignity of Man*. Paul J. Miller, Charles G. Wallis and Douglas Carmichael (eds. and trans.). Indianapolis: Hackett (1998).
Pfohl, Stephen (1990) "Welcome to the Parasite Cafe: Postmodernity as a Social Problem." *Social Problems*, 37:4, pp. 421-442.
Schumacher, E. F. (1977) *A Guide for the Perplexed*. New York: Harper and Row.
Sennett, Richard (1998) *The Corrosion of Character: The Personal Consequences of Work in the New Capitalism*. New York: W W Norton.
Sennett, Richard (2006) *The Culture of the New Capitalism*. New Haven: Yale University Press.
Sparrow, William (2008) "When freaky-deaky equals hara-kiri," Asia Times On-Line: retrieved on 12 October 2008 from <http://www.atimes.com/atimes/Japan/JC08Dh01.html>.

Part III

International Law and Human Rights

8

Climate Change, Developing Countries, and Human Rights: An International Law Perspective

Ved P. Nanda

A broad consensus exists in the scientific community about the reality of climate change. The Intergovernmental Panel on Climate Change (IPCC) has concluded that the concentration of greenhouse gases in the atmosphere has significantly increased globally as a result of human activities in the last 150 years. The more recent appraisal by scientists in several countries is that we may be reaching the "tipping point" and experts warn that climate change could lead to "ecological catastrophes." And the adverse impact will be felt mostly by developing countries without the wherewithal to mitigate these impacts and take the necessary adaptation measures. The international law response has been the 1992 Framework Convention on Climate Change (FCCC) and the subsequent Kyoto Protocol. The Parties to the FCCC failed to reach an accord on a successor treaty at the Copenhagen conference in December 2009. A market-based, flexible approach devised under the Kyoto Protocol to assist the developing countries is the Clean Development Mechanism, which is aimed at reducing carbon emissions and assisting developing countries in achieving sustainable development while allowing advanced industrialized countries flexibility in complying with their emission reduction targets. Several international efforts are underway to link human rights and climate change and to explore legal bases for accountability and state responsibility for global warming.

I. The Challenge

A. Scientific Evidence for Global Warming

A broad scientific consensus exists that climate change is real and has accelerated. In May 2010, the leading scientific organization in the

United States, the National Academy of Science's National Research Council affirmed the reality of climate change, which it said is driven largely by human activities, especially the burning of fossil fuels that release carbon dioxide (CO_2) and other heat-trapping greenhouse gases into the atmosphere, and deforestation (America's Climate Choices 2010). The report, requested by the US Congress, encompassed three comprehensive studies—*Advancing the Science of Climate Change*; *Limiting the Magnitude of Future Climate Change*; and *Adapting to the Impacts of Climate Change*. The requested report

> examines the status of the nation's climate change research efforts and recommends steps to improve and expand current understanding. The report reviews what the scientific community has learned about climate change and its interactions with human and natural systems in 12 areas of interest to decision makers . . . . (National Academy of Sciences 2010, *Advancing the Science of Climate Change, Report in Brief*).

It makes specific recommendations on how to mitigate or adapt to climate change, which include the creation of a carbon pricing system.

In June 2010, the US National Oceanic and Atmospheric Administration (NOAA) released a report, *State of the Climate in 2009*, documenting the weather and climate events in 2009 from around the world and putting them into an "accurate historic perspective, with a particular focus on unusual or anomalous events" (Arndt 2010, S14). The report, compiled by more than 300 scientists from every continent and from more than 160 different research groups, states: "Global average surface and lower-troposphere temperatures during the last three decades have been progressively warmer than all earlier decades, and the 2000s (2000-09) was the warmest decade in the instrumental record" (Arndt 2010, S12). The head of climate monitoring at the United Kingdom's Met Office, one of the collaborating institutions, said: "The whole of the climate system is acting in a way consistent with the effects of greenhouse gases. The fingerprints are clear. The glaringly obvious explanation for this is warming from greenhouse gases" (Harvey July 2010).

Earlier, in November 2007, the Intergovernmental Panel on Climate Change (IPCC), which was established through a collaboration of the United Nations Environment Program (UNEP) and the World Meteorological Organization (WMO), and is comprised of a group of prominent international scientists, unequivocally presented its assessment reports on climate change, concluding that atmospheric concentrations of four long-lived greenhouse gases (GHGs)—carbon dioxide (CO_2), methane (CH_4), nitrous oxide (N_2O), and halocarbons (a group of gases containing fluorine, chlorine, or bromine)—have significantly increased globally as a result of human activities in the last 150 years (IPCC 2007, 37).

The IPCC stated with more than 66 percent assessed probability of occurrence that human activities related to agriculture, fossil fuel use, and land-use are primarily responsible for this change. In the Panel's assessment, advancements since 2001, when the IPCC issued its last assessment report, show that the human impact extends beyond average temperature to temperature extremes and wind patterns, as well (IPCC 2007, 40).

A year later, in October 2008, a report to the government of Australia predicted that CO_2 emissions will continue to rise by more than three percent per year until 2030 (Garnaut 2008). The significance of this prediction is that a growth rate of two percent is the IPCC's median scenario, on which most government projections are based, and its worst-case scenario is that global CO_2 levels would rise by more than two percent per year.

A critical issue to consider in addressing the impact of climate change is how people around the world are likely to be affected. Kemal Dervis, Administrator of the United Nations Development Program (UNDP) and Achim Steiner, UNEP Executive Director, make the point that the poorest and most vulnerable communities have already begun to suffer from its effects. They state in UNDP's *Human Development Report 2007-2008*:

> The effect that increased droughts, extreme weather events, tropical storms and sea level rises will have on large parts of Africa, on many small island states and coastal zones will be inflicted in our lifetimes … [F]or some of the world's poorest people, the consequences could be apocalyptic (UNDP 2008, v).

They add: "In the long run climate change is a massive threat to human development and in some places it is already undermining the international community's efforts to reduce extreme poverty" (UNDP 2008, v).

The UNDP Report's warning is ominous:

> The early warning signs are already visible. Today, we are witnessing at first hand what could be the onset of major human development reversal in our lifetime. Across developing countries, millions of the world's poorest people are already being forced to cope with the impacts of climate change … [I]ncreased exposure to drought, to more intense storms, to floods and environmental stress is holding back the efforts of the world's poor to build a better life for themselves and their children (UNDP 2008, 1).

The Report's unequivocal message is that climate change could lead to "ecological catastrophes" as we are edging toward "tipping points" (UNDP 2008, 2). The outcome could be that the Millennium Development Goals (U.N. Millennium Development Goals 2008) will not be met, which, in effect, means that the world's poor would not be able to satisfy their basic human needs and hence would suffer from widespread violation of the fundamental human rights enshrined in the International Covenant on Economic, Social and Cultural Rights (ICCPR 1966).

In the IPCC's Fourth Assessment Report, which was mentioned earlier, its Working Group II brought to the world's attention the nature of future impacts on developing countries. According to the report, between 75 million and 250 million people in Africa are projected to suffer increased water stress caused by climate change by 2020 (Working Group II, 13). It is also projected that agricultural production, including access to food, will be severely compromised in many African countries, which will adversely affect food security and exacerbate malnutrition. An additional projection is that the sea level will rise, which will affect low-lying coastal areas with large populations (Working Group II, 13). The report further projects that in Asia, glaciers will melt and recede in the Himalayas, which will increase flooding and affect water resources within the next two to three decades. Consequently, more than a billion people could be adversely affected by the 2050s because of the projected decrease of freshwater availability due to climate change, which also will cause an increase in deadly diseases (Working Group II, 13).

As to the impact of global warming on Latin America, the report projects that there will be a risk of significant biodiversity loss in many areas, increased risk of flooding in low-lying areas because of sea level rise, and significant adverse effect on water availability due to changes in precipitation patterns and the disappearance of glaciers (Working Group II, 14). The report notes that small islands located in the tropics or at higher latitudes are especially vulnerable to the effects of climate change, extreme events, and sea level rise (Working Group II, 15). This could affect local resources such as fisheries and exacerbate inundation, storm surges, and erosion; by mid-century, reduction of water resources on many small islands, such as those in the Caribbean and Pacific, will be such that they will become insufficient to meet the population's demand during periods of low rainfall (Working Group II, 15).

B. Appraisal

Scientific evidence for global warming keeps mounting. However, climate change skeptics continue to argue that: temperature measurements are distorted; or while global warming is occurring, it is not attributable to human activities, as was suggested in the various studies mentioned above, and that natural forces are at work, such as solar activity, causing climate change; or global warming has positive consequences, in the form of increased productivity or the increased range of agriculture (Rahmstorf 2004). Many skeptics also accuse some prominent IPCC climatologists of muzzling the voices of opposition and manipulating research, as they point to the leaked emails that surfaced in 2009 that they believe show these scientists to be distorting and hiding data to prove the accuracy of their theories on global warming. They also point to the flaws in the IPCC report on the potential disappearance of the Himalayan glaciers in twenty-five years.

After five different reviews by the British Royal Society, UK House of Commons, the University of East Anglia, in Britain, and Pennsylvania State University of the leaked email exchanges, the accused scientists have been vindicated, as the panels found no reason to dispute the scientists' "rigor and honesty" (McCarthy 2010; *New York Times* 2010). However, the scientists were asked to be more open. The lead scientist at the University of East Anglia's Climate Research unit, Professor Phil Jones, said, "We have maintained all along that our science is honest and sound and this has been vindicated now by three different independent external bodies. There are lessons to be learned from this affair . . . ." (McCarthy 2010).

Although most skeptics have tempered their criticism, skepticism on the warming issue still persists. However, with temperatures on land and sea on the rise, the sea levels also rising, and the declining of the arctic ice sheet—among many other demonstrations of global warming—the scientific evidence on climate change seems to be beyond refute.

II. How Does International Law Address Global Warming?

A. Inadequacy of the Existing Norms of International Environmental Law to Provide the Answer

Two norms that could conceivably apply are Principle 21 of the Stockholm Declaration, which was adopted in 1972 at the UN Conference on the Human Environment in Stockholm, Sweden, and the concept of state responsibility, which was developed by the UN International Law Commission in 1996 in its Draft Articles on State Responsibility. Under these Draft Articles, responsibility does require fault (ILC State Responsibility 1996, 125), that is, a wrongful act or negligence.

Principle 21 of the Stockholm Declaration captures the tension between sovereignty and environmental protection as it states:

> States have, in accordance with the Charter of the United Nations and the principles of international law, the sovereign right to exploit their own resources pursuant to their own environmental policies, and the responsibility to ensure that activities within their jurisdiction or control do not cause damage to the environment of other States or of areas beyond the limits of national jurisdiction (Stockholm Declaration 1972, 3).

The ILC also addressed environmental harm that is unintentional or occurs despite due diligence by establishing a parallel basis for remedies when there is no fault. The ILC called it "state liability" and gave it the title of "Draft Articles on International Liability for Injurious Consequences Arising out of Acts not Prohibited by International Law" [ILC Forty-First Session 1989, 222]. Thus, we have two alternative jurisprudential bases for rectifying harms to the environment—fault-based responsibility, and no-fault (strict or absolute) "liability." The latter means that a state act could give rise to liability even if it did not violate international environmental law.

As we examine the possible application of these norms to climate change, the conclusion is inescapable that neither the Principle 21 approach nor the ILC's liability approach is workable. The "no harm" rule embodied in Principle 21 of the Stockholm Declaration is inadequate to address the climate change problem for several reasons. To start with, it is not easy to trace climate change sources and to measure them, since they are widespread. As it is often states' combined activities that cause climate change, how is one to allocate responsibility among them? Second, while Principle 21 aims at balancing a state's responsibility to avoid harming other nations with its right to exploit its natural resources, most developing nations consider the latter right as their right to economic development according to their own environmental policies, which they consider their priority concern, taking precedence over their abstract responsibility to the international community. Furthermore, the time lag between GHG emissions and their adverse effects makes it impossible to allocate responsibility. In addition, few developing countries have the wherewithal to find alternatives to the fossil fuels on which they are highly dependent. Thus applying the principles of common but differentiated responsibilities and intragenerational equity assumes a central role as we explore the means to respond to climate change. Finally, monetary damages are obviously not an adequate remedy once the damage is done.

The Special Rapporteur of the ILC's Draft Articles on Liability put it well when he stated that the liability approach is premised on state obligations, which presuppose

> an identifiable State of origin, affected State and identifiable harm ... The framework of the topic did not seem to be appropriate for dealing with harm to the human environment as a whole, when there were many States of origin and virtually the whole community of mankind was affected (ILC Fortieth Session 1988, 24).

B. The Alternatives: International Cooperative Measures – IPCC and the Kyoto Protocol and Beyond

Since neither of these approaches can deliver the goods, the focus has to be on international cooperation and prevention. This cooperation is reflected in the UN's efforts to establish a multilateral treaty to address the challenge of climate change. After years of studies, followed by acrimonious disputes between developing and developed countries, with the former blaming the latter for the problem, and after protracted negotiations on an international accord, the UN adopted the Framework Convention on Climate Change (UNFCCC 1992) (Climate Convention) in 1992. The Convention recognized climate change as a serious threat; it set an "ultimate objective [of achieving] stabilization of [GHGs] ... at a level that would prevent dangerous anthropogenic interference with the climate system" (UNFCCC 1992, art. 2). It accepted the principle of differential responsibilities among states, which means that industrial nations would assume greater responsibility to act than developing countries. Although the Convention established a goal of reducing GHG emissions to 1990 levels by the year 2000 (UNFCCC 1992, arts. 4(2)(a), (b)), it provided no concrete targets or timeframe for achieving that goal. Instead, it deferred development of any binding state targets and timetables for a later protocol.

Five years later, in 1997, at the Third Conference of the Parties to the Climate Convention (COP), held in Kyoto, Japan, the parties signed the Kyoto Protocol (Kyoto Protocol 1997), which was created as a framework for future action. The Protocol advanced the implementation process envisaged in the Climate Convention as it included commitments by thirty-seven developed countries and the European Community to reduce GHG emissions, averaging 5.2 percent below the benchmark 1990 concentration levels over the 2008-2012 period (Kyoto Protocol 1997, art. 3(1) Annex B). It also included commitments by the developing countries and introduced market-based "flexibility mechanisms" for implementation, which will be detailed later. The United States did not sign the Protocol because China, India, and other major developing country emitters had failed to make firm commitments.

Selected developments since the Kyoto COP will be highlighted here. In 1998, in Buenos Aires, the Parties adopted a "Plan of Action," setting out a program of work on the operational details of the Kyoto Protocol (UNFCCC 2008). In 2001, the parties adopted the "Bonn Agreements," aimed at completing key issues under the Buenos Aires Plan of Action (UNFCCC 2008), and subsequently at Marrakech (UNFCCC 2002), the signatories to the Climate Convention agreed on rules for implementing the Kyoto Protocol, which came into force on February 15, 2005 (UNFCCC 2005). At Marrakech, a decision was also undertaken to establish an Adaptation Fund "to finance concrete adaptation projects and programmes in developing country Parties that are Parties to the Protocol, as well as [other specifically identified] activities" (UNFCCC 2001, op. para. 1). It was also agreed that the Adaptation Fund "shall be financed from the share of proceeds on the clean development mechanism project activities and other sources of funding" (UNFCCC 2001, op. para. 2).[1] The Fund is governed by a board under the direction of the States Parties to the Protocol.

Further meetings of the COP and the Conference of the Parties, serving as the meeting of the Parties to the Kyoto Protocol (COMP), were held in 2005 in Montreal (UNFCCC COP 2005), in 2006 in Nairobi (UNFCCC COP 2006), and in 2007 in Bali (UNFCCC COP 2007). During the 2007 meeting in Bali, the Bali Roadmap for future negotiations and the Bali Action Plan were adopted (UNFCCC Bali Action Plan 2008, 3).

The Parties took several initiatives at Bali (UNFCCC Bali Action Plan 2008, 3-7). They recognized that "deep cuts in global emissions will be required to achieve the ultimate objective of the Convention and emphasiz[ed] the urgency to address climate change as indicated in the [IPCC's] Fourth Assessment Report" (UNFCCC Bali Action Plan 2008, 3). Key elements of the plan included the launching of a new negotiation process to be completed by the end of 2009 for adoption at the Climate Change Conference in Copenhagen. It is a two-track negotiating process, as the Parties established an Ad-Hoc Working Group on Long-Term Cooperative Action as a subsidiary body under the UNFCCC to conduct the process of negotiating an agreement by 2009 on measures to be undertaken by developed as well as developing country Parties to the Convention. The goal is to establish the Parties' legally binding commitments beyond 2012, when the first commitments to mitigate climate change under the Kyoto Protocol end.

The Parties agreed on a

> shared vision for long-term cooperative action, including a long-term goal for emissions reductions, to achieve the ultimate objective of the Convention ... in particular the principle of common but differentiated responsibilities and respective capabilities, and taking into account social and economic conditions and other relevant factors" (UNFCCC Bali Action Plan 2008, op. para. 1(a)).

On the actions to be considered for mitigation of climate change, the plan contains two separate paragraphs—one for developed country considerations and one for developing countries. For developed countries, the paragraph includes "[m]easurable, reportable and verifiable nationally appropriate mitigation commitments or actions," and for developing countries, "[n]ationally appropriate mitigation actions ... in the context of sustainable development, supported and enabled by technology, financing and capacity-building, in a measurable, reportable and verifiable manner" (UNFCCC Bali Action Plan 2008, op. paras. 1(b)(i)-(ii)).

Along with defining the scope and content of the review of the Kyoto Protocol, the Parties decided to take enhanced action on technology development and transfer, and on the financing of climate change action. The Parties also decided that the Adaptation Fund is to finance concrete adaptation projects and programs that are country-driven. The Parties at Bali took a major decision to include the reduction of emissions from deforestation and forest degradation as one of the considerations during the negotiations to follow Bali (UNFCCC Bali Action Plan 2008, 8, Decision 2/CP.13). The Parties noted that "sustainable reduction in emissions from deforestation and forest degradation in developing countries requires stable and predictable availability of resources" (UNFCCC Bali Action Plan 2008, Preamble). The Program of Work is to be undertaken as related to "a range of policy approaches and positive incentives that aim to reduce emissions from deforestation and forest degradation in developing countries ..." (UNFCCC Bali Action Plan 2008, 9, op. para. 7).

Next, the Parties met in Poznań, Poland, from December 1 to December 12, 2008 to assess the progress since the Bali Conference.[2] The Ad Hoc Working Group, which had met at four sessions during 2008, reported that it had

> considered all of the elements of the Bali Action Plan at each session, taking into account the interlinkages among them ... by addressing a shared vision for long-term cooperative action, enhanced action on adaptation and its associated means of implementation, enhanced action on mitigation and its associated means of implementation, and delivering on technology and financing, including consideration of institutional arrangements (UN FCCC Long-Term Cooperative Action 2008, 2).

At the meeting, the Working Group also reported the ideas and proposals presented by the Parties to it on these elements of the Bali Action Plan (UNFCCC, Long-Term Cooperative Action 2009). The Group's work program for 2009 includes producing a negotiating text in June (UNFCCC Long Term Cooperative Action Work Programme 2009). Among other decisions, the Parties adopted rules of procedure of the Adaptation Fund Board and also its priorities, policies, and guidelines (UNFCCC Adaptation Fund Board 2008).[3] They also adopted the Global Environment Facilities Poznán Strategic Programme on Technology Transfer for Developing Countries, to be funded by €50 million from the UN Global Environment Facility (European Union 2008). They also provided further guidance related to the Clean Development Mechanism (UNFCCC CMP Further guidance 2008).[4] However, delegates made no further progress in advancing the Reducing Emissions from Deforestation and Forest Degradation Plan (REDD 2008).

As the parties met in Copenhagen from December 7 to December 19, 2009, where the negotiation process launched in Bali was to be concluded, the results were disappointing. The Copenhagen Accord (UNFCCC Copenhagen Accord 2009) was "without specific targets or a timetable as to when and how it will translate into a legally binding treaty . . ." (Nanda 2010). The next event was at Bonn—the UN Climate Change Talks—a two-week long session, at the conclusion of which the UNFCCC Executive Secretary stated: "A big step forward is now possible at Cancún, in the form of a full package of operational measures that will allow countries to take faster, stronger action across all areas of climate change" (de Boer 2010). The next UN Climate Change Conference is scheduled from November 29 to December 10, 2010, in Cancún, and will be led by the new UNFCCC Executive Secretary, Christiana Figueres, from Costa Rica.

B. Appraisal

Almost two decades of negotiations since the 1992 United Nations Framework Convention on Climate Change have failed to result in a binding treaty to significantly reduce emissions of greenhouse gases.

The new UN climate change chief, Christiana Figueres, candidly stated in an address to the delegates at the UN Climate Change Conference in Bonn in June 2010: "I do not believe we will ever have a final agreement on climate change, certainly not in my lifetime. [Addressing the issue successfully would] require the sustained effort of those who will be here for the next 20, 30, 40 years" (Harvey June 2010).

What the outcome of the next climate change conference in Mexico, scheduled for December 2010, will be is hard to tell. However, undoubtedly the Kyoto Protocol's response to climate change has not worked, as most countries have fallen short of their targets. And even if they meet those targets, CO_2 emissions will continue to grow. The 2009 economic meltdown and the recession have further called into question the resolve of countries to take aggressive action to cut emissions.

The divide between rich and poor countries persists. Many developing countries are unwilling to commit to targets on reducing emissions without adequate and effective financial assistance and a transfer of appropriate technology on concessional terms from rich countries. Although an agreement was reached at the Bonn meeting on the short-term financing of $30 billion by 2012, and long-term assistance to the tune of $100 billion annually by 2020, unfulfilled pledges of help from rich countries to poor countries in the past have caused distrust.

It may be appropriate to consider alternatives for flexibility in parties' compliance, including the obligations by countries to meet a range of emissions targets determined by their circumstances, and to enable them to opt out on minor issues. Partial agreements on key elements are another possibility. An apt example is the widespread consensus on the Reduced Emissions from Forest Degradation and Deforestation (REDD) program, which was enthusiastically endorsed at the Bonn meeting. In pursuance of the program, Norway has pledged to pay Indonesia $1 billion to preserve large tracts of its forests (Creagh 2010). These variable commitments should be considered in the next phase of the process.

Another alternative is offered by a group of scientists, economists, and policy-makers brought together under the auspices of the London School of Economics and Oxford University. Finding the UNFCCC / Kyoto model as structurally flawed because they felt it misunderstood the nature of climate change as a policy issue, the group presented an alternative plan in May 2010 as a new direction for climate policy in the Hartwell Paper (Hartwell Paper 2010). The group contends that climate change poses a much more complicated challenge than other environmental problems that we have earlier solved. Unlike acid rain or smog, for example, it is not "a conventional environmental 'problem,'" but is as much "an energy problem, an economic-development problem, or a land-use problem" (Hartwell Paper 2010, 15-16).

The Hartwell group proposes the adoption of three climate-related goals: ensuring affordable energy supplies for everyone; ensuring that economic development does not wreak environmental havoc; and ensuring that we are prepared to cope with whatever climate changes might occur (Hartwell Paper 2010, 12-14). It recommends pursuing a number of other worthy goals, such as adaptation, reforestation, encouraging biodiversity, and improving air quality. It seems appropriate to consider the alternatives mentioned here.

III. Addressing the Developing Countries' Concerns and Needs under the Climate Convention and the Kyoto Protocol

A. General

First, developing countries are required to assume mitigation obligations, although the commitment is voluntary, as no specific targets and timetables were set. Under Article 4, paragraph 1, of the Kyoto Protocol, all parties are to establish and report national programs which contain measures to mitigate climate change. Another provision of the Protocol reaffirms this obligation and further seeks to advance the implementation of the developing countries' commitments (UNFCCC 1992, art. 10). Second, the Protocol implicitly recognizes that developing countries are vulnerable to the adverse impacts of climate change as it requires developed countries to provide financial resources and transfer of technology to meet the developing countries costs' of implementing their obligations of emissions reduction (UNFCCC 1992, art. 4(3)).[5]

The Convention established the Global Environment Facility as the financial mechanism to fund developing countries' needs, which is also the entity of the financial mechanism of the Convention operating the Least Developed Countries Fund (Global Development Facility 2008).[6] These Kyoto Protocol provisions reflect application of the principle of common-but-differentiated responsibilities. It should be emphasized that the developing countries' commitments under the Protocol are voluntary and contingent upon the developed countries' assistance. Also, the Adaptation Fund has been established to assist developing countries in their adaptation activities.

B. The Clean Development Mechanism

One of the market-based flexibility approaches devised under the Kyoto Protocol is the Clean Development Mechanism (CDM).[7] The basic elements of the CDM are set out in Article 12 of the Kyoto Protocol and are further supplemented by the 2001 Marrakech Accords (FCCC Marrakech Accords 2002, 20-49), which articulated how this mechanism works. The CDM Executive Board oversees the process under the direction of the States Parties to the Protocol.

The CDM is aimed at reducing carbon emissions. It operates by allowing Annex 1 countries (comprising industrialized countries that were members of the Organization for Economic Cooperation and Development in 1992, as well as countries transitioning from socialist economies, including Russia, the Ukraine, the Baltic States, and several Central and Eastern European states) to earn credits either by governments or private parties in these countries as they engage in project-based activities in developing countries to assist them in reducing their emissions. The credits they earn in developing countries are called "Certified Emissions Reductions" (CERs). CERs are measured in metric tons of CO_2 equivalent and can be sold to buyers in industrialized countries. The CDM Executive Board issues CERs, registers and validates projects, and manages several panels and working groups. Thus, the twin purposes of the CDM are to assist developing countries in achieving sustainable development, and to allow Annex I countries flexibility in complying with their emissions reduction targets.

Electric power plants, wind-based power facilities, and afforestation and reforestation projects that reduce non-CO_2 industrial greenhouse gases illustrate CDM project activities. It should be specially noted that CDM emissions reductions are required to be supplemental to those that would have otherwise occurred without the project, and that a share of the proceeds from certified project activities is to be used to assist developing country parties to meet the cost of adaptation. It is also noteworthy that the only requirement on the part of the host government is that it must affirmatively endorse any CDM project occurring there.

Several potential benefits of the CDM include the reduction of GHGs, technology transfer to developing countries (Seres et al. 2007), and help for developing countries in their adaptation activities, since a percentage of transactions would be targeted for that purpose (UNFCCC 1992, art. 4). The cumulative effect is hoped to result in alleviation of poverty.

Although the CDM program was launched in November 2001, the first project was not registered until three years later, and the first CERs were issued in October 2005 (UNEP Yearbook 2008, 25). As of July 31, 2010, there were 2,307 registered CDM projects in more than sixty countries and more than 4,200 projects are in the registration pipeline (UNFCCC 2010). It is expected that the CDM will generate more than 2.9 billion tradable CERs when the first commitment of the Kyoto Protocol ends in 2012 (UNFCCC 2010, 1). UNEP's Executive Director, Achim Steiner, stated in October 2007 that "\$100 billion of funds are [estimated] to flow from the North to the South as a result of the Clean Development Mechanism" (Steiner 2007, 4).

India (32 percent of registered projects), China (19 percent), and Brazil (13 percent) have dominated the CDM activity (UNEP 2008, 4). A continuing shift in investment from developed to developing countries is in evidence, as the share of new investment has grown from 13 percent (\$1.8 billion) to 23 percent (\$26 billion) in 2007, with China, India, and Brazil together accounting for 82 percent of this investment (UNEP 2008, 4). However, in terms of emission credits generated, China leads with 53 percent, followed by India with just 15 percent. By the end of 2007, \$12.95 billion had been raised by carbon funds (UNEP 2008, 4).

A major challenge is to ensure that countries in Africa benefit from the CDM. To meet this challenge, then Secretary General Kofi Annan launched what is called the Nairobi Framework in 2006. Several UN and affiliated organizations—UNEP, the UN Development Program, the World Bank, African Development Bank, and the UNFCCC Secretariat—came together to implement the Nairobi Framework, with the UN Secretariat acting as catalyst and facilitator (UNFCCC Nairobi Framework 2006).[8] Its initial focus has been to assist six sub-Saharan African countries (Ethiopia, Kenya, Mauritius, Mozambique, Tanzania, and Zambia) in building their capacity to take advantage of the CDM process, with the governments of Spain, Sweden, and Finland contributing \$1.5 million to the project.[9]

Although it has to be a matter of considerable concern that only twenty-seven of the 1,150 registered CDM projects were in Africa as of September 2008, cumulative CDM projects in the pipeline for African countries as of that date were seventy-one, including fifty-one for sub-Saharan Africa. However, as of March 2010, 122 CDM projects were either registered or in the pipeline for validation or registration in Africa (UFCCC Nairobi Framework 2010). The FCCC Secretariat reported in 2008 that "CDM is growing on the continent and is already estimated to be stimulating several billion dollars' worth of capital investment in the seven African countries hosting projects. Market stakeholders and policy-makers are looking for ways to multiply these benefits" (UNFCCC Secretariat 2008). In October 2008, Yvo de Boer, then Executive Secretary of the UNFCCC and the UN's highest-ranking climate change official, exhorted African countries to participate in the current climate change negotiations, which present them "with a golden opportunity to change things for the better and design a Copenhagen deal that works for Africa. For this to happen, it is crucial that African Countries put their concerns on the table and push for solutions that respond to their specific problems" (UNFCCC Secretariat 2008).

C. Other Assistance to Developing Countries

In December 2006, UNEP and the UNDP launched a joint climate change initiative (UNEP Governing Council 2006). This partnership aims at further assisting countries to achieve sustainable development while they confront the changing climate. It extends to all least-developed countries and other developing countries, with a special emphasis on sub-Saharan Africa. Its two core objectives are:

1. Incorporate adaptation into national development plans and UN Cooperation Frameworks.
2. Enable countries to access carbon finance and cleaner technologies to stimulate sustainable development (UNEP Governing Council 2006, 5).[10]

Among several initiatives at the Eleventh Special Session of UNEP's Governing Council, ministers of the environment issued a declaration pledging their governments' commitment "to step up the global response to the major environmental and sustainability challenges of this generation" (UNEP Press Release 2010).

D. Appraisal

As mentioned earlier, the rich/poor divide continues, and there is a lack of trust on whether the rich will provide the necessary assistance to developing countries so that they can take effective steps toward mitigation and adaptation. Among the attempts at the UN, the CDM has come under criticism on its governing practices, environmental integrity, and contribution to sustainable development (Streck 2009). The CDM mechanism is also criticized for not adequately taking into account its projects' adverse impacts upon the human rights of people or on conservation (Orellana 2009). A 2010 study by a UN Human Rights Council expert, Marcos Orellana, adds several more issues pertaining to CDMs for consideration: no requirement of prior informed consent; lack of equitable geographical distribution; equity; failure to promote sustainable development or green technology transfer; lack of access to remedies and jurisdiction; lengthy CDM process; and the lack of transparency (Orellana 2010, 21-23).

Notwithstanding these various criticisms, CDM has been a success story and should continue as a preferable market-based system after 2012. However, remedial action must be taken on the following issues: lack of transparency; lack of access to private parties to challenge executive board decisions; and lack of equitable geographical distribution, along with the lack of a human rights-based approach to CDM.

IV. Linking Climate Change and Human Rights

A. General

Several recent developments link climate change and human rights in light of the growing general awareness leading to broad agreement that climate change will negatively affect the enjoyment of human rights. These include several initiatives at the United Nations, adjudication of cases in various courts, an attempt to invoke the International Court of Justice, and a petition before an international organization.

B. Initiatives at the United Nations

On March 28, 2008, the U.N. Human Rights Council adopted a resolution entitled *Human Rights and Climate Change* (U.N. Human Rights Council 2008), in which the Council requested the Office of the UN High Commissioner for Human Rights (OHCHR) to conduct a "detailed, analytical study on the relationship between climate change and human rights" (U.N. Human Rights Council 2008, para. 1). In connection with the preparation for the requested study, the Australian Climate Justice Program, Climate Action Network Australia, and Friends of the Earth Australia submitted a report to the OHCHR (OHCHR 2008), in which they highlight climate change litigation in several countries.

Earlier, in November 2007, representatives of small island developing states adopted the Malé Declaration on the Human Dimension of Global

Climate Change (Malé Declaration 2007) at a meeting in Malé. They invoked the Universal Declaration of Human Rights, the Stockholm Declaration of the 1972 UN Conference on the Human Environment, the 1992 Rio Declaration on Environment and Development, Agenda 21, the 2002 Johannesburg Declaration on Sustainable Development, and the Plan of Implementation of the World Summit on Sustainable Development, and requested the COP of the UNFCCC to seek the cooperation of the OHCHR and the Human Rights Council in assessing the human rights implications of climate change (Malé Declaration 2007, op. para. 3). They also requested the OHCHR to conduct "a detailed study into the effects of climate change on the full enjoyment of human rights" (Malé Declaration 2007, Op. Para. 4).

In January 2009, the OHCHR released the requested report, in which it noted "broad agreement" on "generally negative effects" of climate change on the realization of human rights (UN Human Rights Council 2009, p. 69). The study detailed implications of climate change for human rights, both directly and indirectly. These were the right to life, adequate food, water, health, adequate housing, and self-determination (UN Human Rights Council 2009, pp. 21-41). It also described in depth the effects of climate change on "those segments of the population who are already in vulnerable situations due to factors such as poverty, gender, age, minority status, and disability," which it listed as women, children, and indigenous peoples (UN Human Rights Council 2009, pp. 42-54). After noting the impact of climate on human migration and on global peace and stability, it referred to some climate change lawsuits to pronounce that "[w]hile climate change has obvious implications for the enjoyment of human rights, it is less obvious whether, and to what extent, such effects can be qualified as human rights violations in a strict legal sense" (footnote omitted) (UN Human Rights Council 2009, p. 70). It provided the following rationale for this pronouncement:

> Qualifying the effects of climate change as human rights violations poses a series of difficulties. First, it is virtually impossible to disentangle the complex causal relationship linking historical greenhouse gas emissions of a particular country with a specific climate change-related effect, let alone with the range of direct and indirect implications for human rights. Second, global warming is often one of several contributing factors to climate change-related effects, such as hurricanes, environmental degradation and water stress. . . . Third, adverse effects of global warming are often projections about future impacts, whereas human rights violations are normally established after the harm has occurred (footnote omitted) (UN Human Rights Council 2009, p. 70).

However, the report did acknowledge that notwithstanding "whether or not climate change effects can be construed as human rights violations, human rights obligations provide important protection to the individuals whose rights are affected by climate change or by measures taken to respond to climate change (UN Human Rights Council 2009, p. 71).

After reciting the states' obligations, the report stated in its conclusions that the effects of climate change "on human rights can be of a direct nature, such as the threat extreme weather events may pose to the right to life, but will often have an indirect and gradual effect on human rights, such as increasing stress on health systems and vulnerabilities related to climate change-induced migration" (UN Human Rights Council 2009, p. 92). Among other conclusions, it stated that while climate change-related harm cannot easily be classified as a human rights violation, "addressing that harm remains a critical human rights concern and obligation under international law" (UN Human Rights Council 2009, p. 96). The report's final conclusion is significant: "International human rights law complements the United Nations Framework Convention on Climate Change by underlining that international cooperation is not only expedient but also a human rights obligation and that its central objective is the realization of human rights" (UN Human Rights Council 2009, p. 99).

At its next session, the Human Rights Council noted the contents of the report and welcomed the decision of its Special Rapporteur on adequate housing to "prepare and present a thematic report on the potential impact of climate change on the right to adequate housing" (UN Human Rights Council March 2009, Op. p. 3).

As the first major study by the U.N. on the linkage between human rights and climate change, the OHCHR report is likely to lead to further studies and discussion in relevant U.N. bodies, especially the Human Rights Council, whose resolution indicates its interest in further pursuing the topic.

C. Lawsuits

In the Nigerian case, *Gbemre v. Shell Petroleum Development Company of Nigeria* (Gbemre v. Shell 2005), a judicial division of the Federal High Court of Nigeria applied several articles of the African Charter on Human and Peoples Rights and the Federal Nigerian Constitution, which includes the right to a "poison-free, pollution-free and healthy environment." The Court held that the gas flaring by Shell and other major oil companies was a gross violation of the applicants' guaranteed rights to life and dignity. The applicants had sought a halt to gas flaring by these companies, which has caused more greenhouse gas emissions in Nigeria than in all other sources in sub-Saharan Africa combined and has also poisoned local communities.

A few U.S. cases will be cited here. In a case before a US federal court, the village of Kivalina, with an approximate population of 400 on the northwest coast of Alaska, filed suit on February 26, 2008, claiming that global warming is destroying the village, which must be relocated or be abandoned soon (Native Village of Kivalina v. ExxonMobil 2008). Relocation will cost hundreds of millions of dollars, and is estimated by the US Army Corps of Engineers and the US Government Accountability Office to range from $95 million to $400 million. The suit is for damages caused by twenty-four energy and utility companies as defendants, alleging that their activities contributed to global warming, which the plaintiffs claim is a nuisance causing severe harm to Kivalina. The defendants' alleged contribution is through their emission of large quantities of GHGs. The plaintiffs stated:

> Each of the defendants knew or should have known of the impacts of their emissions on global warming and on particularly vulnerable communities such as coastal Alaskan villages. Despite this knowledge, defendants continued their substantial contributions to global warming. Additionally, some of the defendants . . . conspired to create a false scientific debate about global warming in order to deceive the public. Further, each defendant has failed promptly and adequately to mitigate the impact of these emissions, placing immediate profit above the need to protect against the harms from global warming (Native Village of Kivalina v. ExxonMobil 2008, para. 5, 2).

The district court for the Northern District of California dismissed the plaintiffs' complaint on political question grounds and as lacking standing. The court concluded that "even accepting the allegations of the Complaint as true and construing them in the light most favorable to Plaintiffs, it is not plausible to state which emissions—emitted by whom and at what time in the last several centuries and at what place in the world—'caused' Plaintiffs' alleged global warming related injuries" (Native Village of Kivalina 2009, 20).

In Comer v. Murphy Oil USA, the United States Court of Appeals for the Fifth Circuit reversed a decision by the federal district court and held that the Plaintiffs, who were residents and owners of lands and property along the Mississippi Gulf Coast, could assert claims against oil, coal, and chemical defendants from property damages resulting from Hurricane Katrina. The Plaintiffs had alleged that the Defendants' operations contributed to global warming through the emissions of GHGs, which caused rising sea levels, thus adding to the intensity of Hurricane Katrina. Subsequently, a majority of the nine Fifth Circuit judges voted in favor of rehearing this case, which was initially decided by a three-judge panel, *en banc*.

The Second Circuit Court of Appeals ruled in Connecticut v. American Electric Power Company, Inc., that two 2004 lawsuits, alleging that GHG emissions by six electric power companies created a public nuisance under federal common law, can proceed against those companies. Subsequently, on March 5, 2010, the Second Circuit denied the petition for rehearing and for rehearing *en banc* filed by American Electric Power Company, et al. (Connecticut v. American Electric Power Company 2009).

There have been some recent efforts to bring climate litigation issues before international tribunals. The tiny state of Tuvalu has tried to institute proceedings against Australia and the United States before the International Court of Justice. The effort is not likely to succeed because not only is there an issue of standing, but the consent requirement is likely to be a major hurdle. If Tuvalu, however, prefers to seek an advisory opinion from the ICJ, the only avenue open is for the UN General Assembly or the Security Council to seek such an opinion. With the potential US veto in the Security Council and the need to have a majority affirmative vote in the General Assembly, Tuvalu's task will not be easy. The other UN organizations authorized to seek an advisory opinion can do so only if the issue falls within their mandate; that will pose an equally formidable hurdle.

D. Petition before an International Organization

The one case where a human rights petition was filed with an international organization is the petition by the Inuit Circumpolar Council (ICC), an international non-governmental organization representing nearly 150,000 Inuit in Alaska, Canada, Greenland, and Chukotka, Russia, before the Inter-American Commission on Human Rights (IACHR) (Inuit Petition 2005). The petition alleged accountability on the United States' part because the US is the world's largest emitter of GHGs. The ICC invoked the human rights of the Inuit as the basis for the petition. Notwithstanding the IACHR's refusal to review the petition, based upon the rationale that "the information provided does not enable us to determine whether the alleged facts tend to characterize a violation of rights protected by the American Declaration" (George 2006)[11], the case has initiated greater awareness about the link between global warming and human rights and the likelihood that international organizations might consider legal bases for accountability and state responsibility for global warming.

IV. Conclusion

Climate change poses a formidable challenge for all countries, but its major impact will be on developing countries, especially the least developed countries, as they lack the resources, capacity, logistics, and wherewithal they need to fulfill their mitigation obligations and to undertake adaptation activities. Thus, the assistance of developed countries becomes imperative. International environmental law, as well as international human rights law, can play a robust role as appropriate mechanisms are crafted to support developing countries in their response to the adverse impacts of climate change. Linking human rights law to the ongoing debate on climate change will greatly assist decision-makers in being informed by a diversity of voices and perspectives on the climate debate and on the range of policies that they consider and ultimately adopt (Hunter 2009). Finally, the matter is urgent. Contrasted with the certainty of climate change is the uncertainty that the international community has the political will to take effective measures to combat it.

Notes

1. Currently, a two percent levy on carbon trading under the UN Clean Development Mechanism finances the Fund.
2. The Parties agreed on a plan of action and programs of work for the year 2008-09 relating to future commitments and actions. UNFCCC, *The United Nations Climate Change Conference in Poznan, COP 14,* November 11, 2008, available at http://unfccc.int/meetings/cop_14/items/4481.php. The second review of the Kyoto Protocol under its Article 9, which requires a periodic review of the treaty in the light of the best available scientific information on climate change and its impacts, as well as pertinent technical, social, and economic information, will also take place there. The capacity of Parties to participate in the CDM will also be reviewed.
3. For the Board's report to the Parties, see UNFCCC CMP (2008).
4. For the Annual report of the Executive Board of the clean development mechanism, see UNFCCC CMP Annual report (2008).
5. Under Article 4(7), implementation of developing countries' commitments depends upon the developed countries' funding and technology transfer.
6. The GEF is designed to fund developing countries' programs and projects that protect the global environment. Its programmatic focus is on: biodiversity, climate change, international waters, land degradation, the ozone layer, and persistent organic pollutants. The Parties provided further guidance for the operation of the Least Developed Countries Fund at the Poznan Conference. See UNFCCC Least Developed Countries Fund (2008).

7. Besides the Bubble Agreement contained in its Article 4, the Kyoto Protocol established three mechanisms for extra-territorial emissions reductions—emissions trading (under Article 17 of the Protocol), joint implementation (under Article 6), and the CDM.
8. The Nairobi Framework has five objectives as key priority targets:
   Build and enhance capacity of [Designated National Authorities] to become fully operational
   Build capacity in developing CDM project activities
   Promote investment opportunities for projects
   Improve information sharing/outreach/exchange of views on activities/education and training
   Inter-agency coordination
   (UNFCCC Nairobi Framework 2006).
9. During the first year of the Framework, Konrad von Ritter, Sector Manager for Sustainable Development at the World Bank Institute, noted:
   There has been notable increase in capacity-development resulting in a pipeline of 30 CDM projects. Of these, 14 have already signed emissions reduction purchasing agreements with World Bank carbon funds. While this is positive we all know that more needs to be done, and therefore the critical importance of the Nairobi Framework to scale up capacity development.
   (UNFCCC 2007)
10. The report makes a telling point:
   To date, the benefits of the Clean Development Mechanism have largely bypassed the Least Developed Countries. Only a handful of countries account for the bulk of registered CDM projects, and there are concerns that the types of CDM projects registered so far provide limited development benefits. To realize the full potential of the CDM as the financing mechanism for sustainable development, a key challenge for developing countries is to remove the institutional, legal and capacity barriers that limit their access to the flourishing and dynamic carbon finance market. To help developing countries address this challenge, UNDP and UNEP will increase their current collaboration in carbon finance, directly supporting the Nairobi Framework on Catalyzing the CDM in Africa agreed at COP12 by six agencies.
   (UNEP Governing Council Appendix 2, 2006:11-12).

References

Arndt, et al, Eds., 2010: State of the Climate in 2009. *Bull. Amer. Meteor. Soc.*, 91 (6), S1–S224.

Comer v. Murphy Oil USA, No. 07-60756 (5th Cir.), October 16, 2009.

Connecticut v. American Electric Power Company, Inc. (2009). Nos. 05-510-CV, 05-5119-CV, Second Circuit, September 21, 2009.

Contribution of Working Group II to the Fourth Assessment Report of the Intergovernmental Panel on Climate Change [Working Group II] (2007), Summary for Policy Makers, *retrieved from* <www.ipcc-wg2.org> on 23 January 2009.

Creagh, Sunanda. 2010. *Indonesia to scrap permits to save forests: official.* Reuters. 31 May 2010, *retrieved from* www.reuters.com/assets/print?aid=USDRE64U15Q20100531, retrieved on 17 June 2010.

de Boer, Yvo. 2010. Press Release. *Bonn climate talks make progress on fleshing out specifics of global climate change regime.* UNFCCC. Bonn, 11 June 2010. *Retrieved from* http://unfccc.int/files/press/news_room/press_releases_and_advisories/application/pdf/20101106_pr_closing_june.pdf on 18 June 2010.

European Union @ United Nations, UN Climate Change Conference: EU Commission Welcomes Poznán Outcome, http://www.europa-eu-un.org/articles/en/article_8368_en.htm (last visited Feb. 12, 2009).

Garnaut, Ross, *Australian Stern Review* (Oct. 27, 2008), *retrieved from* <www.garnau-treview.org.au/index.htm#pdf> on 23 January 2009.

George, Jane. 2006. ICC Climate Change Petition Rejected. *Nunatsiaq News* (2006), *retrieved from* <www.nunatsiaq.com/archives/61215/news/nunavut/61215_02.html> on 25 January 2009.

Global Development Facility (2008), *retrieved from* <www.thegef.org> on 25 January 2009.

Gbemre v. Shell Petroleum Development Corp. 2005. Federal High Court of Nigeria, Benin Judicial Division, Suit no. FHC/B/CS/53/05, *retrieved from* <www.climatelaw.org/cases/case-documents/nigeria/ni-shell-nov05-judgment.pdf> on 25 January 2009.

Hartwell Paper. 2010. *Retrieved from* http://eprints.lse.ac.uk/27939/1/HartwellPaper_English_version.pdf on 25 July 2010.

Harvey, Fiona. June 2010. Further clouded. *Financial Times (London)*. 14 June 2010, 7.

Harvey, Fiona. July 2010. Backing for climate warming findings. *Financial Times (London)*, 29 July 2010, 1.

Hunter, David 2009. Human Rights Implications for Climate Change Negotiations, Oregon Review of International Law 11:331, 340 (2009). International Covenant on Civil and Political Rights (ICCPR), concluded at New York, December 16, 1966, *entered into force* March 23, 1976, 993 U.N.T.S. 171, *reprinted in* 6 I.L.M. 368 (1967).

Inter-American Commission on Human Rights (2005), Inuit Petition: Petition to the Inter-American Commission on Human Rights Seeking Relief from Violations Resulting from Global Warming Caused by Acts and Omissions of the United States, submitted by Sheila Watt-Cloutier, 7 December 2005, *retrieved from* http://www.ciel.org/Publications/ICCPetition_7Dec05.pdf on 22 July 2010.

International Law Commission, Draft Articles on State Responsibility [ILC State Responsibility] (1996), July 12, 1996, art. 1, Report of the ILC on the Work of its Forty-Eighth Session, U.N. Doc. A/51/10 and Corr. 1, 37 I.L.M 440 (1998).

International Law Commission, *Report of the ILC on the Work of its Fortieth Session,* 43 U.N. GAOR, Supp. (No. 10) and U.N. Doc. A/43/10 (1988).

International Law Commission, *Report of the ILC on the Work of its Forty-First Session* [ILC Forty-First Session], U.N. GAOR, 44th Sess., Supp. No. 10, at 222, U.N. Doc. A/44/10 (1989).

Inuit Circumpolar Conference. 2005. Petition to the Inter-American Commission on Human Rights, Seeking Relief from Violations Resulting from Global Warming Caused by Acts and Omissions of the United States, Dec. 7, 2005, *retrieved from* <www.inuitcircumpolar.com/files/uploads/icc-files/FINALPetitionICC.pdf> on 25 January 2009.

IPCC (Intergovernmental Panel on Climate Change) (2007) *Climate Change 2007: Synthesis Report*, *retrieved from* <www.ipcc.ch/pdf/assessment-report/ar4/syr/ar4_syr.pdf> on 23 January 2009. This was the IPCC's Fourth Assessment Report.

Kyoto Protocol to the UN FCCC [Kyoto Protocol] (1997), FCCC Conference of the Parties, 3d Sess., U.N. Doc. FCCC/CP/1997/7/Add.2 (Dec. 10, 1997) (final version), 37 I.L.M. 22 (1998), *retrieved from* www.unfccc.de/resource/confkp.html.

McCarthy, Michael. 2010. "Conspiracy theories finally laid to rest" by report on leaked climate change emails. The Independent (London), 8 July 2010, 8.

Malé Declaration on the Human Dimension of Global Climate Change, adopted 14 November 2007. *Retrieved from* www.ciel.org/Publications/Male_Declaration_Nov07.pdf on 25 July 2010.

Nanda, Ved. 2009. *Not All Lost at Copenhagen Climate Talks,* Denver Post, 24 December 2009, 11B.

National Academy of Sciences, *America's Climate Choices* (2010), *retrieved from* www.americasclimatechoices.org on 31 July, 2010.

Native Village of Kivalina v. ExxonMobil Corp., et al. 2008. Complaint, CV 08-1138 (N.D. Cal., Filed Feb. 26, 2008).

Native Village of Kivalina v. ExxonMobil Corp., et al. CV 08-1138 (N.D. Cal., September 30, 2009).

New York Times. 2010. Editorial: A Climate Change Corrective, 11 July 2010, 7.

OHCHR (2008), *retrieved from* <www.ohchr.org/english/issues/climatechange/docs/submissions/Friends_of_the_Earth_Australia_CANA_ACJP.pdf> on 25 January 2009.

Orellana, Marcos. 2009. A Rights-based Approach to Climate Change Mitigation *in* Greiber, Thomas, Ed. 2009. Conservation with Justice: A Rights-Based Approach, 2009, 37.

Orellana, Marcos. 2010. UN Human Rights Council. Climate Change and the Right to Development: International Cooperation, Financial Arrangements, and the Clean Development Mechanism. UN Doc. A/HRC/15/WG.2/DF/CRP.3/Rev.1, 10 February 2010.

Rahmstorf, Stefan. 2004. *The Climate Sceptics, retrieved from* <http://www.pik-potsdam.de/~stefan/Publications/Other/rahmstorf_climate_sceptics_2004.pdf> on August 1, 2010.

REDD, Envt'l News Service, Climate Hopes and Fears at Poznán, Dec. 5, 2008, available at http://www.ens/newswire.com/ens/dec2008/2008/12/05/01.asp (last visited February 22, 2008).

Seres, Stephen, et al. 2007. *Analysis of Technology Transfer in CDM Projects, retrieved from* <www.cdm.unfccc.int/Reference/Reports/TTreport/report1207.pdf> on 25 January 2009.

Steiner, Achim (2007), *The United Nations Response to the Environmental Challenges of the 21st Century, retrieved from* <www.unep.org/Documents.Multilingual/Default.Print.asp?Documentid=520&articleID=5608&l=en> on 25 January 2009.

Stockholm Declaration of the UN Conference on the Human Environment [Stockholm Declaration] (1972), June 16, 1972, Principle 21, U.N. Doc. A/CONF.48/14/Rev.1 (1973), U.N. Doc. A/CONF.48/14, at 2-65 and Corr. 1 (1972), 11 I.L.M. 1416 (1972).

Streck, Charlotte. 2009. Expectations and Reality of the Clean Development Mechanism: A Climate Finance Instrument between Accusations and Aspirations, *in* Stewart, Richard, et al., Eds. 987 , Climate Finance: Regulatory and Funding Strategies for Climate Change and Global Development. 2009, 67.

UNDP (2008) *Human Development Report 2007-2008 – Fighting Climate Change: Human Solidarity in a Divided World.*

UNEP Governing Council (2006), *Cooperation Between the United Nations Environment Programme and the United Nations Development Programme – Note by the Executive Director,* UNEP/GC/24/INF/19, Dec. 13, 2006.

UNEP Governing Council (2010), *World Environment Ministers Signal Resolve to Realize Sustainable Development*, Bali, Indonesia, 26 February 2010. *Retrieved from* http://unep.org/Documents.Multilingual/Default.Print.asp?DocumentID=612&ArticleID=6482&1=en&t=long on 29 July 2010.

UNEP Governing Council Appendix 2 (2006), *Cooperation Between the United Nations Environment Programme and the United Nations Development Programme – Note by the Executive Director,* UNEP/GC/24/INF/19, Dec. 13, 2006.

UNEP (2008), *Global Trends in Sustainable Energy Investment 2008, Executive Summary, retrieved from* <www.sefi.unep.org/fileadmin/media/sefi/docs/publications/Exec_summary.pdf> on 25 January 2009.

UNEP Yearbook (2008), *An Overview of the Changing Environment.*

UNFCCC (1992), United Nations Framework Convention on Climate Change, May 9, 1992, 31 I.L.M. 849 (1992), *retrieved from* www.unfccc.de/resource/convk> on 25 January 2009.

UNFCCC (2001), FCCC/CP/2001/13/Add.1, Decision 10/CP.7.

UNFCCC (2002), *Report of the Conference of the Parties on its Seventh Session, held at Marrakech from 21 October to 10 November, 2001,* U.N. Doc. FCCC/2001/13 (2002), *retrieved from* <www.unfccc.int/resource.docs/cop7/13.pdf> on 25 January 2009.

UNFCCC (2005), *retrieved from* <www.unfccc.int/meetings/unfccc_calendar/items/2655.php?id=397&out=detail> on 25 January 2009.

UNFCCC (2007), Press Release, *Important Steps Taken to Expand CDM in Africa, Much Remains to be Done – Nairobi Framework Partners,* Dec. 6, 2007, *retrieved from* <www.unfccc.int/files/press/news_room/press_releases_and_advisories/application/pdf/nf_release_english.pdf> on 25 January 2009.

UNFCCC (2008), *Guide to the Climate Change Negotiation Process*, "*Key Landmarks in the Climate Change Process,*" *retrieved from* <www.unfccc.int/resource/process/components/response/landmarks> on 25 January 2009.

UNFCCC Adaptation Fund Board (2008), *Report of the Adaptation Fund Board*, Decision -/CMP.4 (advance unedited version), *retrieved from* <unfccc.int/files/meetings/cop_14/application/pdf/cmp_af.pdf> on 25 January 2009.

UNFCCC Bali Action Plan (2008), Decision 1/CP.13, UNFCCC, *Report of the Conference of the Parties on its Thirteenth Session, held in Bali from 3-15 Dec. 2007 -- Addendum: Decisions Adopted by the Conference of the Parties*, FCCC/CP/2007/6/Add.1 (14 March 2008).

UNFCCC CMP (2008), *Conference of the Parties Serving as the Meeting of the Parties to the Kyoto Protocol, Report of the Adaptation Fund Board -- Note by the Chair of the Adaptation Fund Board,* FCCC/KP/CMP/2008/2 (20 November 2008).

UNFCCC CMP Annual report (2008), Conference of the Parties Serving as the Meeting of the Parties to the Kyoto Protocol, *Annual report of the Executive Board of the clean development mechanism to the Conference of the Parties serving as the meeting of the Parties to the Kyoto Protocol,* FCCC/KP/CMP/2008/4 (14 November 2008).

UNFCCC CMP Further guidance (2008), *Further guidance relating to the clean development mechanism,* Decision -/CMP.4, available at unfccc.int/files/meetings/cop_14/application/pdf/cmp_cdm.pdf.

UNFCCC COP (2005), *retrieved from* <unfccc.int/meetings/cop_11/items/3394.php> on 25 January 2009.

UNFCCC COP (2006), *retrieved from* <unfccc.int/meetings/cop_12/items/3754.php> on 25 January 2009.

UNFCCC COP (2007), *retrieved from* <unfccc.int/meetings/cop_13/items/4049.php> on 25 January 2009.

UNFCCC Executive Board (2008), *Report of the CDM Executive Board for 2007-2008*, FCCC/KP/CMP/2008/4.

UNFCCC Least Developed Countries Fund (2008), Further guidance for the operation of the Least Developed Countries Fund, advance unedited version, Draft decision -/CP.14, *retrieved from* <www.unfccc.int/resource/docs/2008/cop14/eng/02r01.pdf> on 25 January 2009.

UNFCCC Long-Term Cooperative Action (2008), Ad Hoc Working Group on Long-Term Cooperative Action Under the Convention: Report to the Conference of the Parties at Its Fourteenth Session on Progress Made, FCCC/AWGLTA/2008/L.11, Annex (10 December 2008).

UNFCCC Copenhagen Accord (2009), Conference of the Parties, Fifteenth session, Copenhagen, 7-18 December 2009, Draft decision -/ CP.15, FCCC/CP/2009/L.7, 18 December 2009.

UNFCCC Long-Term Cooperative Action (2009), Ad Hoc Working Group on Long-Term Cooperative Action Under the Convention, Ideas and Proposals on Paragraph 1 of the Bali Action Plan, FCCC/AWGLCA/2008/16/ Rev.1 (15 January 2009).

UNFCCC Long Term Cooperative Action Work Programme (2009), Ad Hoc Working Group on Long-Term Cooperative Action Under the Convention, Work Programme for 2009 -- Draft Conclusions Proposed by the Chair, FCCC/AWGLCA/2008/L.10 (10 December 2008).

UNFCCC Marrakech Accords (2002), U.N. Doc. FCCC/CP/2001/13/Add.2/January 21, 2002, at 20-49, *available at* unfccc.int/resource/docs/cop7/13a02.

UNFCCC Nairobi Framework (2006), *retrieved from* <www.cdm.unfccc.int/Nairobi_Framework/index.html> on 25 January 2009.

UNFCCC Nairobi Framework (2010), *Africa source of growth, focus of interest for international emissions offset market*, 5 March 2010, *retrieved from* <http://cdm.unfccc.int/workshops/acf/1003_acf.pdf> on 31 July 2010.

UNFCCC Secretariat (2008), Press Release, *Africa Hardest Hit by Climate Change, Deserves Greater Share of Carbon Market Benefits -- UN's Top Climate Change Official,* Sept. 3, 2008, *retrieved from* <www.unfccc.int/files/press/news_room/press_releases_and_advisories/application/pdf/20080903_africa_carbon_forum_press_release.pdf> on 25 January 2009.

UNFCCC (2010). *CDM Statistics*, 31 July 2010, *retrieved from* http://cdm.unfccc.int./Statistics/index.html on 31 July 2010.

U.N. Human Rights Council (2008). Resolution 7/23, Human Rights and Climate Change, *in* U.N. Report of the Human Rights Council on its Seventh Session, U.N. Doc. A/HR.C/7/78, July 14, 2008, *retrieved from* <www.ap.ohchr.org/documents/E/HRC/resolutions/A_HRC_RES_7_23.pdf> on 25 January 2009.

U.N. Human Rights Council (2009). Report of the OHCHR on the relationship between climate change and human rights, U.N. Doc. A/HRC/10/61, 15 January 2009.

U.N. Human Rights Council (March 2009). Resolution No. 10/4, Human Rights and Climate Change, Report of the Human Rights Council on its Tenth Session, U.N. Doc. A/HRC/10/L.11, 31 March 2009.

U.N. Millennium Development Goals (2008), retrieved from <www.un.org/millenniumgoals> on 25 January 2009.

9

Future Generations' Rights: Linking Intergenerational and Intragenerational Rights in Ecojustice[1]

Laura Westra

This chapter supports the extension of human rights to include the community of humankind and even the community of life that supports it. In general, references to future generations' rights can only be found in the preambular sections of environmental law instruments. The future is viewed as remote, and less worthy of our attention than many other urgent human rights and environmental problems. But the concept of ecojustice the author proposes is intended to join intergenerational and intragenerational justice, and includes the first generation, that is, the generation that is now coming into being and hence is immediately present and fully possessed of rights that we must respect.

Introduction

Climate change is a serious threat to the poor and the vulnerable around the world, especially those in developing countries. It also is a potentially serious threat to future generations. How are we to find appropriate means to ward off the danger to developing countries, especially low-lying coastal states and the least developed states which lack the wherewithal for needed mitigation and adaptation? Equally important, how are we to consider future generations' rights in this context? I suggest that it is essential to link intergenerational and intragenerational rights in ecojustice.

A Philippines Supreme Court case, *Minors Oposa v. Secretary of the Department of Environment and Natural Resources,* affirmed specifically that those inhabiting the Earth today can sue on behalf of future generations (Minors Oposa 1994):

> This case, however, has a special and novel element. Petitioners minors assert that they represent their generation as well as generations yet unborn. We find no difficulty in ruling that they can, for themselves, for others of their generation and for succeeding generations, file a class suit. Their personality to sue on behalf of the succeeding generations can only be based on the concept of intergenerational responsibility insofar as the right to a balanced and healthful ecology is concerned.

This appears to be the only judgment that appeals under international law specifically to intergenerational equity. Barresi goes on to point to the significance of the case: "[I]t was decided by a national court on principles of intergenerational equity for future generations of nationals of that national state" (Barresi 1997, 10). This, I believe, is only partially correct: appeals to future generations for ecological purposes and to preserve "environmental rights," a "nebulous concept" according to Davide, J., who authored the opinion, have far wider implications than the protection of the area's citizens, present and future, as they affect a much larger proportion of the Earth than appears, prima facie, to be the case.

From my point of view, what is particularly important is the appeal to the *parens patriae* doctrine, as the minors request explicitly for "protection by the State in its capacity as parens patriae." As I discuss the rights to health and the environment of children and the preborn, I find the *parens patriae* doctrine to be the best approach to governmental/institutional responsibility for the rights of the first generation. The doctrine has progressed from being used initially for economic/inheritance problems, to juridical use in cases that are exclusively medical and protective; it can also be used for the protection of the lives and health of children and future generations by means of the preservation of a naturally supportive ecology.

Nevertheless, despite its explicit support of intergenerational equity and the novel use of *parens patriae*, subsequent cases did not follow in the footsteps of *Minors Oposa*. In 1997, the courts in Bangladesh took an opposite position in fact (*Farooque v. Government of Bangladesh* 1997).

The major work on intergenerational justice and the law is that of Edith Brown-Weiss. Hence it might be best to approach the topic with a review of the "Sustainable Development Symposium," where she revisits her 1990/1992 argument and responds to the critiques brought against it (Brown-Weiss 1993).

Obligations to Future Generations in the Law: The Proposal of Edith Brown-Weiss

What is new is that now we have the power to change our global environment irreversibly, with profoundly damaging effects on the robustness and integrity of the planet and the heritage that we pass on to future generations.

What are the main characteristics of Brown-Weiss's position? The first thing to note is that her proposal comprises both rights and duties, and that these include both "intragenerational" and "intergenerational" aspects. Intergenerational duties include the obligations:

1. to pass on the Earth to the next generation in as good a condition as it was when that generation first received it; and
2. a duty to repair any damage caused by any failure of previous generations to do the same.

Thus every generation has the right "to inherit the Earth in a condition comparable to that enjoyed by previous generations" (Barresi 1997, 2). In addition, each generation has four duties:

1. conserve the diversity of the Earth's natural and cultural resource base;
2. conserve environmental quality so that the Earth may be passed on to the next generation in as good a condition as it was when it was received by the present generation;
3. provide all members with equitable access to the resource base inherited from past generations; and
4. conserve this equitable access for future generations.

These duties impose non-derogable obligations, especially on affluent Western developed countries, which are clearly in a position of power, as most of the degradation, disintegrity, elimination of biotic capital, and other serious ecological ills proceed directly from the practices of the powerful West, to the vulnerable South. I have argued that these obligations should be viewed as *erga omnes*, and they should also be considered as founded on *jus cogens* norms, as the proliferation of harmful chemicals, the exploitation of natural areas, and the many activities exacerbating global climate change represent a form of institutionalized ecological violence, or ecoviolence, on vulnerable populations. As gross breaches of human rights, they should be thus considered to be ecocrimes and treated accordingly.

In contrast, some have argued that both limitations on economic expansion and commercial activities on one hand, and the demand for increased respect for the preservation of endangered areas and species on the other, only represent a Western, imperialistic conceit, one that flies in the face of the South's needs and cultural practices. Guha and others contrast the Western concern with the environment as a source of leisure-time amenities, rather than understanding its role as foundational to survival, as has been demonstrated by many, including the World Health Organization (WHO) (Guha 1989, 312-319).

This partial understanding allows Guha to make a specious distinction between humanity and its habitat, something that is biologically impossible. WHO's research has demonstrated the impossibility of separating human health and normal function from environmental conditions, and the consequences of human technological activities in regard to children and the preborn. Nevertheless, it is obviously true that it is easier for developed countries to institute remedial regimes to correct and restore presently harmful environmental conditions, than it is for developing countries to do so.

Thus Brown-Weiss is quite correct as she links intergenerational obligations with intragenerational duties: rich countries and groups must discharge their duties intergenerationally in a direct form, but also fulfill their intragenerational obligations to developing countries and impoverished populations. The latter would not be able to fulfill their own obligations without help. But the institutions of rich countries can, and therefore must ensure that the global communal obligations to future generations be met not only by them, but also by those who require their help in order to comply (this relation between the ability to help and the duty to do so can be found in the "Kew Garden Principle" of 1972) (Velasquez 2000).

The principle of "equitable resource use" can therefore be understood in this way: rather than exacerbating a conflict between North-West preferences and South-East basic needs, as Guha proposed, combining the two—under the Kew Garden Principle—ensures that both intergenerational and intragenerational basic rights are met and the correlative obligations are discharged. Paul Barresi lists Brown-Weiss's proposed rights and duties, and her strategies for the implementation of these duties. He acknowledges that her point is that these should be more than just moral obligations: they should be codified as law. Strategies of implementation include establishing a Planetary Rights Commission, which might serve as a forum where individuals and groups might bring complaints for the violations of these environmental rights. Perhaps courts might be set up to complement the International Criminal Court

(ICC) and other ad-hoc tribunals, intended to bring to justice serious human rights violations.

I have argued that ecocrimes represent gross breaches of human rights and should be judged accordingly, and no less seriously than attacks against the human person, genocide, breaches of global justice, and crimes against humanity in general (Westra 2004). Thus appealing to international criminal law might even eventually allow a Planetary Rights Commission to be part of the International Criminal Court. But far more important than thinking of the courts appropriate to try and convict those guilty of these crimes against future generations, is to prevent these irreversible harms from happening, by ensuring now that ecojustice should prevail by supporting it in both its aspects of intragenerational and intergenerational equity. Both these aspects should be codified in appropriate law regimes, and both should be enforced.

An Aside on Justice and Ecojustice: Rationality in Natural Law

> Il exist un droit universal et immuable, source de toutes les lois positive; il n'est que la raison naturelle, entant qu'elle gouverne tous les hommes. (*Code Napoléon*)

Ross notes the existence of at least three traditions of justice, basic to understand its full meaning: the positivist doctrine, where each country's "sovereign" expresses his will through the law, which therefore has an indisputable binding force, and the natural law theory, which views justice as an a priori rational principle. Ross adds that, therefore, the "enacted law" in this case possesses binding force only to the extent to which it is a realization or attempt at a realization of the idea of law. Finally, there is "the romantic or historical school of law," where law is custom, or "the mirror of the popular mind" (Ross 1958, 106-107).

The first alternative does not exist in today's legal regime, but both positive law and a principled approach may be found in common and civil law respectively, or they may both be present in some proportion, as they are, for instance, where "general principles of law," including natural law and customary law, are explicitly considered as part of the sources of international law.

Justice may also be understood as the highest virtue, including its reach in both law and morality. Justice as a principle and a concept permits us to assess the cogency and the thoroughness of laws and judicial decisions by how they compare to that standard: "As a principle of law, justice delimits and harmonizes the conflicting desires, claims and interests in the social life of the people ... Justice is equality" (Ross 1958, 268).

Of course, perfect equality is not a viable principle, nor well-founded, and thus justice may take into consideration merit, performance, needs, or ability. We need not elaborate on issues that are very well discussed in the literature, particularly those arguments raised by John Rawls. What should be emphasized, however, is that certain human characteristics may never be taken into consideration if justice is to be served: these include race, gender, creed, and social class. John Rawls's "difference principle" captures well the need to ensure that the least advantaged be treated in a way that alleviates their disadvantage in someway, and that corrects, as does the law, the unfairness of their situation.

The second approach to justice that Ross proposes, that is, natural law, is well adapted to provide the foundation for this understanding of justice, even if one prefers to purge it of its metaphysical underpinnings, as Hugo Grotius suggests, saying that, even if one assumed that God did not exist, "the law of nature would still be valid." I have argued that, as natural law provides the basis for just war theory, it is also foundational to the understanding of crimes against humanity, genocide, and other breaches of basic human rights that singly and collectively represent aspects of what I have termed "ecoviolence."

That said, just as certain characteristics of humankind such as color, gender, or creed should not be used to dictate a discriminatory treatment of certain humans, or the deprivation of their right to protection from harm, so too the geographical or temporal distance of humans from the actor or decision-maker ought not to permit discriminatory treatment either. The argument in favor of rationality as foundational to justice, even without further inquiry about the origin of such rationality, can be found in Kantian doctrine, and it has been used to great advantage by such eminent Kantians as Onora O'Neill and Thomas Pogge on the topic of global justice (O'Neill 1996).

The main theme of global justice according to these thinkers is the continued presence of starvation and poverty globally, despite the rhetoric of providing aid, on the part of Western affluent countries. For instance, the 1996 World Food Summit in Rome issued a pledge by the 186 participating governments that ends with the following words: "We consider it intolerable that more than 800 million people throughout the world, and particularly in developing countries, do not have enough food to meet their basic nutritional needs. This situation is unacceptable" (Rome Declarations on World Food Security).

This situation is also exacerbated and fostered by global climate change, and we have witnessed the lack of interest in signing on to the

Kyoto Agreement by the country that is the worst polluter: the United States. Yet the International Monetary Fund (IMF), the World Bank, and the World Trade Organization (WTO), led by the United States, "had unprecedented power to shape the global economic order." The World Bank recognizes that global poverty will increase, not decrease, reaching 1.9 billion people by 2015.

Pogge discusses the failures of these institutions by showing "three morally significant connections between us and the global poor." In brief, these are the causal/historical connections of a shared past where the presently wealthy imposed colonialism and slavery on the presently poor, so that the foundation of the former's joint powers and affluence is suspect. Another connection is that we all depend on a single "resource basis"; we share the Earth, but only in ways that benefit those better off and harm the others. Finally, the present "global economic order" does not redress, but aggravates "global economic inequalities" (Pogge 2001, 14).

The difficulty Pogge emphasizes through the example of global famine supports Richard Falk's contention that the world as a whole has arrived at a "Grotian moment," but thus far we have been unable or unwilling to even attempt to reach across the "normative abyss" that is before us: "A neo-liberal world order based on the functional imperatives of the market is not likely to be a Grotian moment in the normative sense" (Falk 1998, 14).

His main point coincides with the argument presented here: while Grotius was able to "articulate a normative bridge between past and future" at a critical historical moment, the present world order is both unable and unwilling to do so.

Falk notes that the present world order "reflects mainly economistic priorities," and that the state system has lost much of its credibility in problem solving in the face of "the rise of market forces." Falk adds:

> We currently confront in this era of economic and cultural globalization a more profound normative vacuum: the dominating logic of the market in a world of greatly uneven social, economic, and political conditions and without any built-in reliable means to ensure that a continuing global economic growth does not at some point and in certain respects cause decisive ecological damage (Falk 1998, 26).

Neither Pogge nor Falk addresses intergenerational justice directly, and some who do so, like Hendrik Ph. Visser 't Hooft, find Rawls's work to be more helpful than I have found it to be. In sum, Rawls refuses to commit to a specific formulation of "the good" as superior to other possible choices, as he allows perhaps too much power to the "rational contractors," even when their formulation of just principles takes place "behind a veil of ignorance." In contrast, his "difference principle" might be extended, as Visser 't Hooft would have it, to global society (Visser 't Hooft 1999, 55-56).

In general, accepting the arguments for sustainability entails that moral judgments be used when scientific evidence is assessed and legal instruments are often designed to support such judgments:

> Justice constitutes one segment of morality primarily concerned not with individual conduct but with the ways in which classes of individuals are treated. It is this which gives justice its special relevance in the criticism of law and other public or social institutions (Visser 't Hooft 1999, 54).

From the environmental point of view, intergenerational justice emerges in several conventions, such as the Biodiversity Convention, as biodiversity, for instance, is acknowledged to be a "common concern of mankind," but it cannot be said that there are in any convention substantive provisions to treat living resources as a "common heritage" or give full effect to intergenerational rights as conceived by Brown-Weiss.

Some additional examples of this somewhat limited concern, or of the presence of human interests, are left to the preambles of international treaties. For instance, the Preamble to the Whaling Convention (1946) recognizes "the interests of the actions of the world in safeguarding for future generations the great natural resources represented by whale stocks"; the 1968 African Convention, states that soil, water, and fauna resources constitute "a capital of vital importance for mankind"; the 1985 ASEAN Agreement, talks about "the importance of natural resources for present and future generations"; the 1972 World Heritage Convention says that "parts of the natural heritage are of outstanding interest and therefore need to be preserved as part of the world heritage of mankind as a whole"; and finally the Preamble to the 1973 Convention on International Trade in Endangered Species (CITES) speaks of wild flora and fauna as "an irreplaceable part of the natural systems of the earth which must be protected for this and future generations to come" (Birnie and Boyle 2002, 605).

Even such a brief discussion indicates that while Brown-Weiss's work considers both intergenerational and intragenerational equity, most existing legal instruments do not. I have suggested that the integration of the two concepts represents true ecojustice, that is, justice that recognizes humans as embedded in their habitat, so that justice that does not recognize this aspect of humanity is—to say the least—incomplete. Other internationally used concepts, such as "sustainable development," are mostly used as a device to attempt to "green" an existing or projected policy or proposal that simply advances "business as usual."

It is clear that justice as a normative, rational principle is well-served by Brown-Weiss's doctrine in three ways. First, it requires that present populations needing help to achieve appropriate environmental goals should receive such help from the affluent countries who, it must be noted, have more than a moral interest in ensuring the conservation of common resources. Next, it avoids critiques based on identifying specific individuals to ensure that we should understand fully what their needs (or preferences) might be in the future: it is not individual rights, but group/class rights that are at issue. All considerations of equity in regard to certain groups and classes look at the general characteristics of all, NOT at the individual preferences of any specific member of the group, as some have argued against the rights of future generations. Finally, it defends implicitly the right of the present generation to protection that accrues from the present respect for "nature's services." Thus, instead of showing an unfair bias toward the future, at the expense of the needs of the poor of the present generation, it mandates the preservation and the protective instruments that are needed for all present peoples, especially the most vulnerable, such as the poor, those in developing countries, and the children. All these require sustainability in their habitat, which Visser 't Hooft terms, "the ecological core of the concept, which after all is the historical prior one, and ... the judgments on quality of life called for by the conception of a future integrity of the environment" (Visser 't Hooft 1999, 17).

When so understood, Brown-Weiss's doctrine on intergenerational justice not only reconciles intergenerational justice and intragenerational justice, but it also manifests the underlying integrity and unity of justice and ecojustice. Paul Wood argues in a similar vein by addressing the question of "The Connection Between Intergenerational Equity and Allocational Equity." He proposes designing specific regimes to meet the joint goals of developed and developing countries, especially those in deltaic regions or island countries: "An equitable self-sustaining regime will meet intertemporal obligations by jointly benefiting present and future generations" (Wood 1996, 293-307).

Visser 't Hooft also remarks that intergenerational justice involves classes of individuals, but in a very novel, original way. I believe he is mistaken on this point. Decisions proscribing apartheid, for instance, or discrimination against people of color, once codified in law, are equally intended to cover unknown groups of individuals, included in an open-ended class of persons fitting, for example, the definition of "persons of color." Those legal instruments are based on principles equally "processional and open-ended" as future generations are. He adds: "Justice between generations forms a natural alliance with the search for transgenerationally valid criteria of environmental integrity, established by means offering the best chances for consensus and clear definition" (Visser 't Hooft 1999, 54).

In contrast, I have argued, following James Karr, Reed Noss, and other biologists and ecologists, that ecological integrity needs a scientific understanding and an index (IBI) to allow the reconstruction of a baseline. Such a definition is certainly scientific and thorough, but it is not based on consensus. I have also proposed that justice, both transgenerational and intergenerational, would best be served by adherence to a "principle of integrity" that would morally prohibit any activity or process that involved humanmade products such as the ones that are putting at risk the most vulnerable among us, starting with the children of the first generation, but extending globally to the present poor and all future generations as well.

Klaus Bosselmann on Ecological Justice

Klaus Bosselmann has noted that there is dissonance between most environmental ethics theories, which do not really address social justice issues, and theories of social justice that do not fully appreciate the impacts of ecological problems. His analysis of the problem starts by noting that "a theory of either environmental justice or eco-justice is lacking." He cites a definition of environmental justice that views it as "equal justice and equal protection under the law without discrimination," but he also points out that such a view ignores the intergenerational aspect of the concept (Bosselmann 1999).

My work on environmental racism is admittedly also focused primarily on "Africana," that is, issues of concern to North America's African American population, and hence it is guilty of the same narrow scope for the most part.

Nevertheless, it is in the environmental ethics work that I have addressed future generations' rights, not the work on environmental justice. In the latter, it was important to note the impact of then President Bill Clinton's Executive Order No. 12,898 (Federal Actions to Address Environmental Justice in Minority Populations and Low-Income Populations, 11 Feb. 1994), which ensured that the U.S. Environmental Protection

Agency (EPA) would be more available to redress environmental injustices and harms befalling the inhabitants of African-American neighborhoods, than they had been before the introduction of the Executive Order. The grave concern for the life and health of presently affected populations had clear primacy over the possible rights of the future.

Yet, it must be admitted, a consideration of the latter would have eliminated the need for costly clean-up and remediation on behalf of the former. In addition, there is a problem in that minority groups and peoples from developing countries tend to perceive ecocentrism as a position in direct conflict with their own, justified, immediate needs and aspirations. Bosselmann acknowledges that this position portrays a false dichotomy: "The concentric position is inclusive, as it merely extends intrinsic values of humans to non-humans rather than replacing one by the other" (Bosselmann 1999, 34).

I have argued in this way as well in defense of the "Principle of Integrity," as the debate between anthropocentrism and eco/biocentric holism is based on misconceptions about the scientific underpinnings of the latter position, the holistic biocentrism I have defended. The two positions are seldom, if ever, combined or even discussed together in the literature, although I have argued that they converge in the interface between health, normal function, and ecological integrity.

Bosselmann argues for an approach to "eco-justice," based on Brenda Almond's proposal, involving:

- the relation between modern liberal theory of justice and environmental ethics;
- the various forms of distributive justice with respect to the environment; and
- their application to environmental issues (Bosselmann 1999, 37).

This set of guidelines appears to be basically flawed. There are, of course, grave difficulties produced by relying on neoliberal policies. Bosselmann himself sees the wrongheadedness of this approach:

> But like the "rights" issue, the liberal approach of justice tends to foster the very problems we are trying to overcome. Rather than fundamentally challenging the traditional idea of environmental management with its anthropocentric limitations, it would simply internalize the concept in the "idea of environmental justice" (Bosselman 1999, 39).

Finally, invoking distributive justice leads to the problems on which Richard De George, for instance, bases his argument against future generations' rights, as well as introducing the difficulties present in various forms of "exponential discounting" (De George 1981, 289).

Bosselmann wants to link intragenerational justice and intergenerational justice, citing Brown-Weiss's own proposal noted above and extending the meaning of "future generations" to non-human animals. I have proposed going even beyond that, by including all life under the same protective umbrella, thus including the unborn, or first generation as well. By starting with the consideration of health and normal function, thus relying not only on ecology, but also on epidemiology and the work of the WHO, the form of ecojustice proposed here is indeed radical. But, by connecting existing regulatory regimes not only to their explicit environmental—even if non-anthropocentric—thrust, but also to their implicit interface with all human health, I believe this proposal for ecojustice might be the most extensive one, best suited to inform supranational and international law regimes. In the next section, the case of the First Nation Peoples of Canada will demonstrate the links between intergenerational and intragenerational harms, connected through health considerations.

Intergenerational Harms: The Case of the Canadian First Nation Children

> Chiefs must consider the impact of their decisions on the seventh generation (paraphrase of precept of the Great Law of the Haundenosaunee [Six Nations Iroquois Confederacy]; Seventh Generation Fund, P.O. Box 4569, Arcata, CA 95518).

In a case out of Winnipeg, Manitoba, Canada, only "environmental" in an extended sense, the abuse of a common chemical compound, glue, by an expectant mother, posed a grave threat to the health of the yet unborn child. She had made similar choices previously, to the irreversible detriment of her two other children. Aside from the question of the possible restraints of the addicted mother, there are two important issues that also emerge from the case, and that are particularly relevant in a discussion of group rights and transgenerational and intergenerational equity.

The first point of significance is the use of the *parens patriae* doctrine which, as we will see, is here also used for the protection of group rights. The second point relates directly to specific harms suffered by aboriginal peoples, especially the harmful legacy initiated by the policy of "assimilation" of First Nation children. Social science research shows today a dismal picture of harm to these children, as the following statistics indicate:

Child Functioning by Aboriginal and non-Aboriginal Children in Care

- Substance Abuse Birth Defect: Aboriginal Children (7%), non-Aboriginal Children(4%), Other visible minority children (1%)
- Behavior Problem: Aboriginal Children (18%), non-Aboriginal Children (25%); Other Visible Minority Children (18%)
- Irregular School Attendance: Aboriginal Children (15%); non-Aboriginal Children (10%), Other Visible Minority Children (6%) (Blackstock, Trocme, 2004, Table 4).

This table, coupled with what is presently known about pre-birth exposures, indicates a persistent, ongoing, intergenerational form of injustice. Unchecked, this injustice leads routinely to harms to all First Nation's Peoples, as most of these children's problems arise because of exposures and circumstances beyond their control.

Many of the difficulties these children face are due to earlier Canadian and Provincial policies of "assimilation," intended to eliminate all forms of cultural "Indian-ness" from children, by placing them in white institutional schools, removed from their parents and extended families except for two months each Summer, forbidden to speak their native tongues, and forced to abandon all their traditional practices.

These practices, intended to educate Aboriginal children to "the ways of the white man," without any respect or appreciation for their own traditions and culture, placed a heavy burden on the children, and produced harms that cannot be righted, even with the best will, within one generation. The results of earlier Canadian policies continue to produce "intergenerational effects" today:

> Aboriginal communities have not yet recovered from the damage caused by the residential schools ... For the first time in over 100 years, many families are experiencing a generation of children who live with their parents until their teens (Report of the Aboriginal Justice Inquiry of Manitoba: the Justice System and Aboriginal People 1991).

This is not the appropriate place for a lengthy historical discussion of the failures of the Canadian Government policies in regard to its Aboriginal Peoples. The main point is that this is another case where intergenerational justice should have been practiced in several areas:

> (1) pre-birth protection, in order to prevent many of the childhood problems statistically prevalent in Aboriginal children; (2) the proper implementation of *parens patriae* doctrine in the defense and protection of Aboriginal children with behavioral, emotional, and neurological problems arising (in part) from environmental or other substance exposure; and (3) the serious

and informed consideration of intragenerational issues connected with the intergenerational ongoing harm inflicted upon these children.

Starting with the final point as the most obvious one, Canada's Aboriginal Action Plan addresses the question of "intergenerational effects":

> The intent of the residential school policy was to erase Aboriginal identity by separating generations of children from their families, suppressing their Aboriginal languages, and re-socializing them according to the norms of non-Aboriginal society (ibid.).

As noted earlier in the discussion of Brown-Weiss's proposal, one of the major objections to her position was the open-endedness of the proposed obligation to the future. This is an example of harms to a whole class, present, future, and even to the past generation; it is the reality of all social and legal changes imposed on Aboriginal children.

The first point mentioned above is also of greater importance to First Nation Children than to others, given the statistics exposing the grave problem disproportionately affecting them because of pre-birth drug and alcohol exposure. These problems affect children immediately, but they also add to delinquency in older children and adults, thus representing another form of intergenerational injustice. Against this background, however, Canada's Child Welfare Statutes give primacy to the "best interests of the child."

> Although these documents start with the "Respect for family autonomy and support of families," and add "Respect for cultural heritage especially for Aboriginal Children," they also include "the paramountcy of the protection of children from harm." Certainly, protection of children from harm should not only initiate after their birth, when the endangerment could happen long before it. Health Canada produced a study, "Canadian Incidence Study of Reported Abuse and Neglect." Nicholas Bala writes: "Each child welfare statute has a definition of a "child in need of protection" (or an "endangered child"). This is a key legal concept as only children within this definition are subject to an involuntary state intervention under the legislation" (Bala 2004, 19).

If the need for protection arises long before social workers or the courts may be confronted with it, then it is worthy of serious consideration at the time of exposure, that is, at the time of the origin of the harm. In other words, it implies that pre-birth child endangerment should be viewed as impermissible and that legal and social measures should be put in place before the harm occurs. Legal instruments are needed to prevent such harms before they affect both the child and society. Serving justice by preventing the harms would make both logical and economic sense as well. As the greater percentage of affected children is Aboriginal, it would seem that any instrument that does not consider the pre-conditions of their humanity is flawed and incomplete.

Intergenerational and intragenerational justice would seem to require this additional protection, as additional protection seems to be necessary, for example, from the freedom of chemical companies and other polluters to impose harm on the most vulnerable, based on the extensive research of the WHO.

Finally, the second point, the implementation of the *parens patriae* doctrine, appears to be as justified to support the intervention required by the first point as it was in the case where, for instance, a Jehovah's Witness parent refused a blood transfusion to the serious detriment of the child. As Justice MacLachlin said:

> Para. 76. Canadian Child protection Law has undergone a significant evolution over the past decades. This evolution reflects a variety of policy shifts and orientations, as society has sought the most appropriate means of protecting children from harm. Over the last 40 years or so, society has become much more aware of problems such as battered child syndrome and child sexual abuse, leading to a call for greater preventive intervention and protection. At the same time, Canadian law has increasingly emphasized individual rights to protection against state intervention (R.B. v. Children's Aid Society of Metro Toronto 1995).

In addition to the Jehovah's Witness case cited above, La Forest, J. discusses the implications of that decision regarding s.7 of the Charter (citing Whealy Dist. Ct. J.):

> When an infant, totally incapable of making any decision, is in a life threatening situation and the appropriate treatment is denied or refused by its parents, it cannot be said that any potential protection as given under section 7 for the family unit can be invoked against the right of the infant to live. Section 7 addresses itself also to "the principles of fundamental justice." It can hardly be said that the principles of fundamental justice could be invoked to deny a child a chance to live. It is worth noting as well that the rights set out in section 7 are conditional and not absolute. The rights therein set out can be interfered with if done in accordance with the principles of fundamental justice. The scheme of the Child Welfare Act, in my view, meets all the tests of fundamental justice, including a fair hearing before an impartial judge (para. 54) (ibid.).

It is important to note two major points here. The first is that the freedom of religion was here appealed to by the parents of Sheena B., the infant at risk, as they attempted to block the medically required blood transfusion. Freedom of religion is a very important principle, far superior, I believe, to the freedom of choice to engage in a lifestyle that includes the use of alcohol or drugs. Neither is protected by the Charter.

The second point is the appeal to "principles of fundamental justice": if they could not be invoked "to deny a child the chance to live," I believe they also could not be invoked to deny to a child the right to be well-born, that is, to be born with normal physical and mental function.

The Rights of the First Generation and of the Future: The Interface

> [B]y its very nature, law must play a protective role. If violence and dependency to strong men in a society is to be replaced by the rule of law, law fulfills a role according to its most fundamental principle, i.e., to protect the equal dignity of all men, in freedom and mutual responsibility (Hirsch, Ballin Ernest, M.H. 1999, 7).

Children are the world's citizens. But while they are children they cannot speak on their own behalf or represent themselves, and one cannot always guess exactly what their future choices and preferences might be. These are also the characteristics of future generations, in fact, the very characteristics that render future generations' rights hard to defend both in morality and in the law. I have argued that ecoviolence, an attack against the human person perpetrated through environmental means, is an accepted and institutionalized criminal activity against an "other" we do not want to recognize as worthy of respect. These "others" could be citizens in developing countries, where the ecoviolence acquires a sinister racist perspective, as well as poor and "different" citizens of the most developed countries. The "other" may also be one of our own: the developing child of this population, that is, the first of future generations.

Similar thoughts and sentiments are found in an unexpected area of scholarship: in the reflections on the Holocaust. Eva Hoffman says: "Systematic violence—especially what Primo Levi called "unnecessary violence"—that is the violence that does not serve the ends of battle or victory but is meant to humiliate and brutalize the victim, is the ultimate form of mis-recognition or deliberate non-recognition" (Hoffman 2002, 280).

The "meaning," the "intent," is absent from environmental harms, and also lacking in pre-birth and perhaps some other harms to children. But, as "deliberate cruelty" is judged to be an attempt to discount, negate, and ultimately destroy the identity and the subjectivity of its target, perhaps we can also say that the converse is true. In other words, "to discount, negate the subjectivity" of those who are harmed, even without the requisite *mens rea*, necessary for the commission of a crime, is ultimately, a form of destruction that manifests an unthinking but undeniable cruelty. The humanity of these "others" is kept faceless, their identity denied.

Is this possible account of the converging plight of the first and future generations overly dramatic? Perhaps. But if we cast our minds back, to consider the consequences of our refusal to recognize the mindless violence offered to future generations and to the unborn through environmental means, we will have to acknowledge that such wholesale harms are not compatible with respect for humanity. Even within a war situation, where institutionalized violence is expected and accepted, it is not the case that any "externality" (in capitalist/consumerist terms), or "collateral damage" (in just war terms) is considered to be acceptable either morally or legally.

In the ambit of just war, there are still *jus in bello* conditions to be met, even when the objective of winning a battle, or a just war itself, is at issue. Senseless cruelty or violence is not condoned, nor is any attack against medical personnel, prisoners of war, hospitals, or the cultural or religious icons of a people (Geneva Convention 1948). These restrictions apply even when the goal to be achieved is a legal one and—for the most part—morally just. Therefore, it seems fair to also ensure that both corporate profit-making enterprises, and the activities of common citizens, whose lifestyles may impose grave "collateral damage" on those who are negligently or unintentionally harmed, be subject to similar stringent restrictions.

The freedom of persons, be they legal or biological, is not absolute and many restrictions are already in place: through public health (e.g., freedom of movement restrictions because of communicable diseases, from measles to T.B.; of smoking in public places; of having unprotected sex when HIV positive as in the recent Canadian murder conviction); or through police powers (e.g., through restrictions of speed of movement on highways, or the very impermissibility of driving if intoxicated or under the influence of certain substances).

In these cases and many others, liberty is limited by our responsibility not to harm others, a responsibility that is not only moral, but also legal and enforced through public institutions and the courts. One may object that the harms imposed by drunk driving or second-hand smoke, or even TB exposure, are clear and obvious. However, based on the scientific research of the WHO and of many epidemiologists, the results of unrestrained activities by industry or single individuals may soon well be considered "clear and obvious" too. In some cases, these effects may not be immediately visible, but neither is the result of second-hand smoke, HIV exposure, global warming, or even drunk driving. The latter, for instance, may continue for a period without accident; but the potential for accidents and grave harm increases exponentially when the harmful preconditions persist.

Of course, neither the harm to the unborn nor to the next generation may be immediately observable. But the scientific research today is clear and robust enough to prompt us to adopt laws that defend the rights of the future.

Edith Brown-Weiss and Others on the Protection of the Future: *Standards of Fairness*

> *Standards of Fairness.* Experts generally adopt one of two general approaches to fairness. One measure depends on a priori rules, or "focal points"—cultural or ethical rules broadly perceived as measuring fairness, which derive authority from legal and cultural traditions. The second approach views fairness as the outcome of strategic bargaining emerging from the negotiating process instead of being pre-determined. Bargained for rules derive their legitimacy from the explicit consent of agreeing parties (Brown-Weiss, quoted in Wood 1996, 305).

Given that the latter is usually subject to the will of Western power blocs as a main guiding principle, and that negotiated treaties are usually brought down to the lowest possible denominator, I have argued that ethical norms, including natural law, represent a far better option to ensure that justice and fairness be served. In fact, although *erga omnes* obligations do not explicitly include the rights of future generations, it seems that the right to survival and, if only minimally, to the health and normal function that will make such survival meaningful and re-affirm the basic rights of future people, are and should be at least as important as the right to escape racist or gender-based harms.

Hence, *jus cogens* norms, not politically motivated agreements, should govern regulatory regimes to protect the survival of the future. Treaties support the obligations of signatories, that is, of certain specific like-minded countries, but the harms of which we have been speaking are global in reach. Brown-Weiss speaks of the "planetary legacy" that cannot be abused or consumed, as it presently is, through the "produce of Western technologies and lifestyles." Wood suggests the creation of "cooperative regimes" shifting "the focus away from benefits, costs and targets," as in the present Climate Change Protocol. He also proposes that Western developed countries should provide financial and technical assistance to developing countries, many of which are becoming major polluters, such as China, for instance.

But Wood himself recognizes that, if standing for future interests is not universal, protection would be geographically as well as temporally incomplete. In that case, relying on voluntary agreements and regimes would appear to be precarious at best, and to represent an approach that is doomed to failure. In contrast, Brown-Weiss views intergenerational obligations and the intragenerational duties that support them, as *erga omnes* in character. This appears to be the best available option, as neither negotiated treaties, nor a "world-based empathy for the environment," could possibly be used to counteract the world hegemony of the U.S. and the WTO, and their indisputable ability to dominate global policy decisions they appear to view exclusively through an economic lens.

Brown-Weiss's position is convincing instead: the protection of future generations is specifically mentioned in various international instruments, and in fact, the World Commission on Environment and Development (WCED) recommended the appointment of an ombudsman for future generations. In addition, treating the "Planetary Trust" appropriately means that, like it, the case for all trusts is that "maintenance of the capital is an integral component of the investment." In general, trusts require the long-term perspective demanded by the consideration of future generations. Finally, the recognition of the rights of the future is but another aspect of a human right, not an anomaly, or a different issue altogether. The Preamble to the Universal Declaration of Human Rights begins as follows:

> Whereas recognition of the inherent dignity and of the equal and inalienable rights of all members of the human family is the foundation of freedom, justice and peace in the world (UDHR 1948).

Considering the language of this Declaration, Brown-Weiss concludes that,

> the reference to all members of the human family has a temporal dimension which brings all generations within its scope. The reference to equal and inalienable rights, affirms the basic equality of such generations in the human family (Brown-Weiss 1992, 21).

Both the UN Declaration and Brown-Weiss's interpretation support the understanding of the rights of the future as imposing an *erga omnes* obligation on all presently existing members of the human family in respect to the future. The non-derogability of this obligation shows that neither preferences nor negotiations may eliminate this grave duty.

Therefore Wood's argument for viewing, for instance, the climate system as "a common property resource," though well-intentioned, misses the full understanding of the situation. The "commons" will be the topic of the next section.

The International Protection of Human Rights and the Principle of the Common Heritage of Mankind

> In our search for an Ariadne's thread to lead us through the intricacies of international relatives we stumble upon a new concept creeping in and out of the intricacies of international reality: the "common heritage of mankind" (Cassese 1989, 376).

When we consider the historical development of international law, we note that the powerful Western countries have tended as much as possible to support the status quo in law, through the respect for the "free will of states" and the prevalence of custom and of positive law. In contrast, the developing countries and Eastern Europe, albeit for separate motives, supported the formation of principles and rules beyond those based on the agreement and cooperation of states. Cassese's insightful analysis shows that, while in earlier years a "Hobbesian or realist tradition" prevailed, which saw each country's position as essentially self-defensive in regard to other states, a later Grotian or internationalist conception of state interaction emerged, emphasizing "cooperation and regulated intercourse among sovereign States" (Cassese 1989, 31). Finally, the universalist "Kantian outlook" emerged, "which sees at work in international politics a potential community of mankind, and lays stress on the element of 'transnational solidarity'" (Cassese 1989, 31). The latter approach, together with a strong thrust toward the emergence of *jus cogens* norms and of obligations *erga omnes,* represent the preferred approach of developing countries as they press "for quick, far reaching and radical modifications" (Cassese 1989, 123).

In contrast, Eastern European countries "prefer to proceed gingerly, believing as they do, that legal change should be brought about gradually, as much as possible through mutual agreement" (Cassese 1989, 123).

Nevertheless, both Eastern European and developing countries joined in supporting Article 53 of the Vienna Convention on the Law of Treaties (1969). Cassese remarks:

> To developing countries, the proclamation of *jus cogens* represented a further means of fighting against colonial or former colonial countries—as was made clear in 1968 at the Vienna Conference by the representative of Sierra Leone, who pointed out that the upholding of *jus cogens* provided a golden opportunity to condemn imperialism,

> slavery, forced labour and all the practices that violated the principle of equality of all human beings and of the sovereign equality of all states (Cassese 1989, 176).

For Eastern European states (such as Romania and Ukraine), *jus cogens* was viewed as a "means of crystallizing once and for all, peaceful coexistence between East and West."

Despite the support of both of these blocs, Western countries were initially on the defensive before bowing to the inevitable will of the majority, and the need to espouse norms consistent with their own legal traditions. It is instructive to consider that it is the weakest countries, those which felt most disempowered by Western alliances and treaties, which enthusiastically supported a "new law" approach, characterized by the introduction of *jus cogens* norms (although the "old law"—the Westphalian order where only states were the subjects of international law—also took a similar position). E. Jimenez de Arechaga (Uruguay) termed these developments "a flagrant challenge to international conscience" (Cassese 1989, 178). I have argued that we are all disempowered in the face of mounting environmental threats to our health and survival, and the powers that support global trade and current economic policies instead of life (Westra 2006, 157). Perhaps that is why we see protesting groups joining forces not only from developing countries, but also from Western environmental and animal defense groups. At any rate, Cassese summarizes the three principles that were codified in the 1969 Vienna Convention: "First, it introduces restrictions of the previously unfettered freedom of States; Second, there is a democratization of international legal relations; Third, the Convention enhances international values as opposed to national claims" (Cassese 1989, 189).

A controversial principle is arising within the "new law" paradigm, one that has not quite lived up to its true potential, at least so far: the Principle of the Common Heritage of Mankind. The concept appears *prima facie* to step forward, but this does not represent the whole picture. Birnie and Boyle say:

> An important factor contributing to the classification of living resources as common property is that they have generally been so plentiful that the cost of asserting and defending exclusive rights exceeds the advantages to be gained. A regime of open access in these circumstances has generally been to everyone's advantage. However, as Hardin has observed, the inherent logic of the commons remorselessly generates tragedy, as the availability of a free resource leads to overexploitation and minimizes the interest of any individual state in conservation and restraint (Birnie and Boyle 1992, 118).

The Principle of the Common Heritage of Mankind

Law does not spring anew; old concepts evolve and new ones emerge to fit new fields of human enterprise. In this manner, the unique historical developments manifesting themselves in the emergence of a North-South cleavage have been responsible for the introduction of a new international legal concept: the Common Heritage of Mankind (CHM) Principle. The new legal concept, CHM, can be defined as follows: the area under consideration cannot be subject to appropriation, all countries must share in the management of the region, there must be an active sharing of the benefits reaped from the exploitation of the area's resources, and the area must be dedicated to exclusively peaceful purpose.

This appears, at least *prima facie*, to be a wonderful addition to the small arsenal of ecologically constructive concepts. Nevertheless, the language employed in that definition shows clearly its incompleteness and deficiencies. If an area is ecologically sensitive and important enough to fit the CHM concept, then both managing it and exploiting it may be contrary to the continued preservation of the area, a goal implicit in the CHM designation. All future generations comprising humankind would be deprived of any benefit whatsoever if the area were to be both managed and exploited.

That goal would be far better served if both present and future humankind were managed instead, so that their exploitive activities could be controlled and even excluded from the area to be designated as a common heritage: the area's existence and the natural services it may provide for all life both within and without its immediate confines, are what is primarily at stake.

In essence, future generations of humankind can only benefit from non-exploitation, which, in turn, is based on regulated restraint or management of present human enterprise. Although the CHM principle is not yet established as either a treaty obligation or as an obligation *erga omnes,* it remains a "political principle" at this time and has emerged in international discourse because developing countries have been seeking a New International Economic Order (NIEO).

The developing nations, largely disempowered by free trade and the economically and politically powerful G-8, are attempting in this way to influence public policy opinion, at least in regard to areas "outside the traditional jurisdiction of states: the deep seabed, outer space and, to a lesser degree, the Antarctic" (Larschan and Brennan 1983, 310).

If one is concerned with the national systems of Earth, the focus has to be on the deep seabed and Antarctica. The first point worthy of note in this regard is that this political principle is too accepting of the status quo, and hence is not capable of protecting our common heritage as stated, because this natural patrimony of humankind does not only lie in areas that do not interest the Northwest affluent states. Oceans, old forests, lakes, rivers, and all other areas where biodiversity still abounds are surely part of the global commons and should be protected urgently, before their tragic loss may deprive all life of the support they provide.

Hence, the reference to the "benefits" of exploitation is clearly an oxymoron, unless one interprets benefit in a purely economic and short-term sense that appears to be contrary to the letter and spirit of a principle aimed at benefiting humankind as a whole, not only a rich and present minority.

Are we, for instance, to consider the global commons as *res nullius*, despite the tragic consequences that may follow the free and unrestricted appropriation of these areas by technologically advanced countries and other legal persons, bent on immediate economic exploitation? Or are we to consider it *res communis*, together with air and sunlight? The 1974 separate opinion of Judge De Castro in the case on the *Fisheries Jurisdiction* shows clearly the fallacy of this approach, as he states that "fish stocks in the sea are inexhaustible." But neither clean air nor safe sunlight is presently available to most people on Earth, and fish stocks themselves are often sadly depleted or have crashed into extinction (Westra 1998, Chapter 6).

I contend that the Common Heritage of Mankind Principle should be applied as *territorium extra commercium*, as Bin Cheng proposes, except that instead of "management, exploitation and distribution," our concern should be with preservation, non-manipulation, and respectful treatment, as these concepts, not the former, would ensure that humankind as such may enjoy the benefits of an unspoiled nature.

Here is Cheng's important passage discussing some of these concepts:

> While territorium extra commercium and Territorium commune humanitatis (for CHM) shared the same characteristics that they cannot be territorially appropriated by any State, they differ, in that the former is essentially a negative concept, whereas the latter is a positive one. In the former, in time of peace, as long as a State respects the exclusive quasiterritorial jurisdiction of other states over their own ships, aircraft and spacecraft, general international law allows it to use the area or even abuse it more or less as it wishes, including the appropriation of its natural resources, closing large ports of such ports of such space for weapon testing and military exercises

> and even using such areas as a cesspool for its municipal and industrial sewage. The emergent concept of the common heritage of mankind, on the other hand, while it still lacks precise definition, wishes basically to convey the idea that the management, exploitation and distribution of the natural resources of the area in question are matters to be decided by the international community (or simply the contracting parties, as in the Moon Treaty?) and are not to be left to the initiative and discretion of individual States and their nationals (Cheng 1980, 337).

Larschan and Brennan, in contrast, are primarily concerned with distributive issues. They argue convincingly that, even though it seems that certain areas are protected under the CHM Principle, in practice it only appears to protect the "Group of 77," given that the "one nation-one vote procedure of the Assembly is cosmetic." The Council empowered to make executive decisions is dominated by states "on the basis of investments, social system, consumption, production, special interests and equitable geographical distribution" (Larschan and Brennan 1983, 323).

Our concern, instead, is with long-term preservation, not with the present distribution of the economic benefits of the global commons. The distributive approach, as Cheng points out, permits the use of the patrimony of humankind as a "cesspool," hardly appropriate to the Common Heritage of Mankind. Even with its weaknesses, it would have been highly desirable to retain the use of the principle beyond open space, the moon, and the deep seabed. The Antarctic Treaty System protects the area and the related ecosystems "in the interest of mankind as a whole" (1991 Protocol to the Antarctic Treaty on Environmental Protection, Preamble), and the most obvious common heritage—the air we breathe—has not been so designated, rather the "global climate" has been referred to as a "common concern." In other environmentally related preambles, the expression used is "world heritage of mankind" (Convention for the Protection of World Cultural and Natural Heritage 1972).

In contrast, referring to any aspect of the commons as a "property resource" eliminates the requirement of respect and preservation and substitutes an approach that only requires "fairness of allocation procedures." The latter will not support ecojustice as the combination of intragenerational and intergenerational justice, and replaces it with an approach that retains an economic/procedural flavor, as "maintaining quality, allocating capacity, and controlling access." I have argued that we are facing the "final enclosure movement" as the tragedy of the commons reaches its final stage.

To propose, as Wood does, that "stable institutions [that] include equitable arrangements, efficiency, assumed expectations through compliance monitoring and graduated sanctions" (Wood 1996, 311), shows a misunderstanding of both the nature and the gravity of the situation. It is not only a matter of slowing down the inevitable elimination of the resource base by procedural fairness and "assurance games." It is rather a question of viewing the Earth not only as property to be divided and exploited, even when fairness is employed, but to consider it as comprising natural systems whose integrity and support are essential to our survival present and future.

The protection of ecosystemic functions supports our own, as the WHO has indicated, both in the present and in the future, starting with the first generation.

Intragenerational and Intergenerational Equity: Ecojustice for the First and for Distant Generations

> This we know: the earth does not belong to man: man belongs to the earth ... Whatever befalls the earth befalls the sons of the earth. Man did not weave the web of life: he is merely a strand in it. Whatever he does to the web, he does to himself (Chief Seattle, patriarch of The Duwamish and Squamish Indians of Puget Sound, to U.S. President Franklin Pierce in 1855).

This is indeed the position of today's ecology, and of biocentric and ecocentric environmental ethics, from Aldo Leopold on. There are both scientific and moral reasons to support the need for ecojustice, or equity that respects the future as well as the present. From a scientific point of view, the unpredictability of future events, based on recent chaos theory research, ensures that any prediction that makes claims to certainty and accuracy is most likely incoherent and false. Scientific uncertainty is an accepted paradigm today, but even the use of the precautionary principle is, in some sense, insufficient, as it promotes the idea that we are not sure whether ecological or biological harm will follow certain practices or activities. In contrast, what might be uncertain or imprecise might be the specific form the expected harm will have, not its occurrence. In this sense, it might be like saying that devices capable of predicting the occurrence of a tsunami (such as the one that devastated Indonesia, Sri-Lanka, and Thailand on Dec. 26, 2004) should not be put in place, because such devices cannot predict exactly the number of victims for each affected country, or the precise amount of economic damage we can expect.

The precautionary principle proposes that we should err on the side of caution, because we are not sure. But many of the harms resulting from ecological/biological disintegrity are well—if not precisely—known and expected. The problem is not lack of knowledge, but a combination of stubborn partiality for short-term gain, and for visible immediate advantages, particularly economic ones, over both precaution and long-term safety, and for the consumerist/capitalist thrust of corporate activities. Moreover, these latter activities are insulated and protected against the undefended rights of vulnerable peoples and populations to survive unharmed.

That is why such soft law instruments as the Earth Charter are widely praised and welcomed in the developing world, but viewed with suspicion and distrust by wealthy, affluent countries, although the latter are the ones most in need of its principles of respect for all generations, and for the integrity of the Earth. Morally speaking, Brown-Weiss is certainly correct as she affirms, "As part of the natural system, we have no right to destroy its integrity, nor is it in our interest to do so. Rather, as the most sentient of living creatures, we have a special responsibility to care for the planet" (Brown-Weiss 1990, 198).

The combination of up-to-date science and moral belief culminates in some of the legal instruments cited earlier. The "right" sentiments are often expressed in preambles and such, but the quest for true equity escapes: economic interests tend to block normative considerations. Thus, harmful exposures destroy the lives and decimate the healthy functioning of the first generation, while harmful substances continue to accumulate to wreak a worse havoc on the future.

A link between first and future generations can also be found in the economic concerns found in most families. At least two common approaches found in families favors the future: first, the accumulation of wealth on the part of many parents in consideration of the future needs of their children, is useful "both as insurance for the parents, and an inheritance for the children" (Epstein 1989,1465). Second, the motive to secure a better future for the children prompts parents to reduce spending in favor of saving: "the bequest motive thus tends to defer consumption and promote investment" (Epstein 1989, 1472-1473).

This example illustrates some of the consequences of the serious concern for the future we may find in parents, but it is probably not readily present in the general population in relation to our own collective future offspring. Both the desire not to waste, but to respect (ecological) wealth, and to curb consumption of (natural) resources, would provide an excellent basis for the intergenerational equity we are seeking to promote. The aspect of respect is certainly present in the beliefs and attitudes of aboriginal peoples, including the First Nation peoples of Canada, and most African peoples.

Respect, in turn, may well breed concern and even distaste for overconsumption, when the consequences that will surely follow are clearly understood. But neither respect nor restraint may be left to spring up spontaneously in the hearts and the thoughts of all people, although both are natural and expected when our own future (first) generation is under consideration. That benevolence must be legislated and enforced if we ourselves and our future are to be protected and if equity is to be required.

The Significance of Equity to Ecojustice

In his important Dissenting Opinion in the Gabcikovo-Nagymaros Case, Judge Christopher Weeramantry discusses equity in detail. Perhaps most important, in his long and thorough analysis, Weeramantry says that "in the context of sharing of natural resources ... equity is playing an increasingly important international role" (Gabcikovo-Nagymaros Case 1997).

Equity, as Weeramantry explains it, is far more than procedural fairness. It may be used by the courts as rationale to reach a decision when faced by facts that may not have been considered before in the Court's jurisprudence; but equity, as Judge Jimenez de Arechaga affirms, cannot be used to reach a "capricious" decision, but must exhibit "reasonableness in the light of individual circumstances." It can be used in the sense of "applying distributive justice and redistribution of wealth," should equity considerations permit, and in fact mandate, the consideration of the ensuing results of a judicial decision. Finally, it is intended "to render justice [not?] through the rigid application of general rules and principles of formal legal concepts, but through an adaptation and adjustment of such principles, rules and concepts to the facts, realities and circumstances of each case" (ibid. 124).

Thus, the arguments purporting to prove that no full knowledge or precision is available to direct and inform our thinking when seeking a just approach to all future generations, fail. Equity itself may be brought in as the necessary corrective, through its role in "tempering the application of strict rules" (ibid.). Weeramantry cites Aristotle in the Nichomachean Ethics in support of his position:

> The reason for this is that law is always a general statement, yet there are cases which it is not possible to cover in a general statement ... This is the essential nature of the equitable: it is the rectification of law where law is defective because of its generality (ibid. 133).

But the judicial leeway and flexibility here indicated show that judicial discretion (c) can only be applied through the choice of an equitable principle.

As noted above, not only are some of the causes of the harms to which all future generations are exposed, from the first one onward, similar, but also the reasons why these exposures are not yet clearly proscribed in the law have similar roots. They include the belief in absolute freedom as paramount for both individuals and legal persons, and the preeminence of the economic motive—the so-called "sovereignty" of the consumer.

It is hard, in fact almost impossible, to restrain freedom and preference satisfaction even in the name of intragenerational equity; it is much harder to do so for intergenerational motives. Nevertheless, it should be possible to do so, as Weeramantry's dissent suggests, because many principles of equity are already embedded in the law: "Many principles of equity such as unjust enrichment, good faith, contractual fairness and the use of one's property so as not to cause damage to others are already embedded in positive law. In the field of international law, the position is the same" (ibid. 115).

The appropriate use of equity, in fact, starts with assessing which facts and circumstances must be considered. The existence of the harms befalling the future from present practices, it seems clear, cannot and should not be excluded from any consideration used to reach an equitable position regarding those practices.

Conclusion

States as well as human beings, by both their action and inaction, share the responsibility for causing emissions and building greenhouse gases (GHGs) that have both intergenerational and intragenerational impacts. Several authors in this book have eloquently articulated the need for a renewed environmental ethic as we face global climate change. Equitable considerations enshrined in the "common but differentiated responsibilities" principle lie at the core of the legal and institutional mechanisms devised under the Kyoto Framework as the world community strives to undertake the needed action to respond to climate change (Nanda, this volume).

How the post-Kyoto scenario unfolds at the Copenhagen meeting in December 2009, scheduled to unveil the successor to the Kyoto Protocol, with fixed obligations and timetables for each state, will demonstrate whether there is in fact the political will to realize the urgency of securing both intergenerational and intragenerational rights.

Note

1. This article is based on a chapter previously published in the author's *Environmental Justice and the Rights of Ecological Refugees*, Earthscan 2009.

References

Adams, Hussein M. (1995) "Somalia Environmental Degradation and Environmental Racism," in L. Westra and P. Wenz (eds.), *Faces of Environmental Racism*. Lanham, MD: Rowman Littlefield.

Bala, Nicholas (2004) *Canadian Child Welfare Law*. Bala et al. (ads.), 2nd edn.

Barresi, Paul A. (1997) "Beyond Fairness to Future Generations: An Intergenerational Alternative to Intergenerational Equity in the Intergenerational Environmental Arena," *Tul. Envt.L.J.*

Birnie and Boyle (1992) *International Law & The Environment*, Oxford University Press, Oxford.

Birnie, Patricia and Adam Boyle (2002) *International Law and the Environment*. 2nd edn., Oxford University Press.

Black's Law Dictionary (1979).

Blackstock C. and Trocme N. (2004), "Pathways to the Overrepresentation of Aboriginal Children in Canada's Child Welfare System," Social Services Review, Vol.10.

Bosselmann, Klaus (1999) "Justice and the Environment: a Theory on Ecological Justice," *Environmental Justice and Market Mechanisms*, Bosselmann and Richardson (eds.), London: Kluwer Law International.

Brown-Weiss, Edith (1988) "The Planetary Trust: Conservation and Intergenerational Equity," *Ecology L.Q.*

Brown-Weiss, Edith (1993) "Intergenerational Equity: Toward an International Legal Framework," in Nazli Chourcri (ed.), *Global Accord*, MIT Press, Cambridge MA.

Brown-Weiss, Edith (ed.) (1992) *Environmental Change and International Law*. Tokyo: United Nations University Press.

Brown-Weiss, Edith (1990) "On Rights and Obligations to Future Generations for the Environment," *Am. J. Int'l L.*

Cassese, Antonio (1989), *International Law in a Divided World*, Clarendon Press, Oxford.

Cheng, Bin (1980), "The Legal Regime of Airspace and Outer Space in the Boundary Problem, Functionalism versus Spatialism: The Major Promises," *Annals of Airspace Law*.

Child and Family Services Act, s.1(a), as amended S.O. (1999) c.6.

Child, Family and Community Services Act, R.S.B.C. (1996) c.46, s.2(a) and 4(1).

Code Napoleon, cited in Ross, Alf (1958) *On Law and Justice*. Stevens and Sons Limited.

Daily, Gretchen (1997) "Introduction," in *Nature's Services*. Washington, DC: Island Press. D'Amato, Anthony (1990) "Agora: What Obligation do Our Generation Owe to the Next? An Approach to Global Environmental Responsibility," *Am. J. Int'l L.*

D'Amato, Anthony (1992) "Agora: What Obligation do Our Generation Owe to the Next? An Approach to Global Environmental Responsibility," *Am. J. Int'l L.* De George, R. (1981) "The Environment, Rights and Future Generations," in E. Partridge (ed.), *Responsibilities to Future Generations*. Buffalo, NY: Prometheus Books.

Epstein, Richard A. (1989) "Justice Across the Generations." *Tex.L.Rev.*

Falk (1998), *Law in an Emerging Global Village*, Transnational Publishers, Ardsley NY.

Farber, Daniel A. (n.d.) "From Here to Eternity: Environmental Law and Future Generations," p. 289; Farber suggests that discounting is unavoidable, but he proposes 'the perpetuation value of environmental resources and the use of Hyperbolic discounting where the discount rate itself declines over time.'

Farooque v. Government of Bangladesh (1997) 49 DLR (AD) 1.

Gbadegesin, Segun (2001) "Multinational Corporations, Developed Nations and Environmental Racism: Toxic Waste, Oil Exploration and Ecocatastrophy," in L.Westra and H. Lawson (eds.), *Faces of Environmental Racism*, op.cit.

Goedhuis (1981) "Some Recent Trends in the Interpretation and the Implementation of International Space Law." *Colum. J. Transna'l L.*

Goerner, Sally J. (1994) *Chaos and Evolving Ecological Universe*. Amsterdam, The Netherlands: Gordon and Breach Science Publishers.

Grotius, Hugo [edition] (1625) *De Jure belli ac pacis*, libri tres.

Guha, Ramachandra (1989) "Radical Environmentalism and Wilderness Preservation: A Third World Critique," in L. Pojman (ed.), *Environmental Ethics*. 4th edn., Belmont, CA: Wadsworth Publishing.

Hart, H.L.A. (1961) *The Concept of Law*. Oxford: Clarendon Press.

Hirsch, Ballin Ernest, M.H. (1999) "Children as World Citizens," in *Globalization of Child Law*. The Netherlands: Martinus Nijhoff Publishers Dordrecht.

Hoffman, Eva (2002) "The Balm of Recognition; Rectifying Wrong Through the Generations," in *Human Rights, Human Wrongs*. Oxford, UK: Amnesty International Lectures.

Human Resources Development (2000) *Child Welfare in Canada*. Canada.

Karr, James & Ellen Chu (1995) "Ecological Integrity: Reclaiming Lost Connections," in L. Westra and J. Lemons (eds.), *Perspectives on Ecological Integrity*. Dordrecht, The Netherlands: Kluwer Academic Publishers.

Koskenniemi, Martti (1992) "Breach of Treaty or Non-Compliance? Reflections on the Enforcement of the Montreal Protocol." *YB Int'l Envtl. Law*, 3.

Larschan, Bradley and Bonnie C. Brennan (1983) "The Common Heritage of Mankind Principle in International Law." *Colum. J. Transnat'l L.*, 21.

Leopold, A1do (1949) *A Sand Country Almanac and Sketches Here and There*. New York: Oxford University Press.

Maritime Delimitation in the Area Between Greenland and Jan Mayer (Denmark v. Norway) (1993) I.C.J. 38.

McMichae1, A.J. (2000) "Global Environmental Change in the Coming Century: How Sustainable Are Recent Health Gains?" in D. Pimentel, L.Westra and R. Noss, *Ecological Integrity: Integrating Environment, Conservation and Health*. Washington, DC: Island Press.

McMichae1, A.J. (1995a) *Planetary Overload*. UK: Cambridge University Press.

Minors Oposa v. Secretary of the Department of Environment and Rural Resources, 33 I.L.M. 173 (1994), Davide, J. R.J.

Norton, Bryan (1995), "Why I am not a Nonanthropocentrist: Callicott and the Failure of Monistic Inherentism," in *Enviornmental Ethics,* Vol.17 Winter.

Noss, Reed F. (1992) "The Wildlands Project: Land Conservation Strategy." *Wild Earth*, Special Issue.

Oke, Yemi (2005) "Intergenerational Sustainability and Traditional Knowledge in Africa," paper presented at the Ecological Integrity Conference, Venice, Italy, July 2, 2005.

O'Neill, Onora (1996) *Towards Justice and Virtue*. Cambridge, MA: Cambridge University Press.

Ottawa (2002) retrieved from <http://www.hrdc.gc.ca/sp-ps/socialp-psociale/cfs/rpt2000/rpt 2000e_toc.shmtl>.

Parfit, Derek (1984) *Reasons and Persons*. Oxford, UK: Oxford University Press.
Partridge, Ernest (n.d.) "On the Rights of Future Generations," in D. Scherer (ed.), *Upstream/Downstream*. Philadelphia, PA: Temple University Press.
Pimentel, D., L. Westra and R. Noss (2000) *Ecological Integrity: Integrating, Environment, Conservation and Health*. Washington, DC: Island Press.
Pogge, Thomas (2001) *Global Justice*. Oxford, UK: Blackwell Publishers.
Protocol to the Framework Convention on Climate Change (Kyoto) (1998) I.L.M., 37, 22.
R.B. v. Children's Aid Society of Metro Toronto (1995) 1 S.C.R. 315.
Rawls, John (1971) *A Theory of Justice*. Cambridge, MA: Harvard University Press.
Rees, William (2000) "Patch Disturbance, Ecofootprints, and Biological Integrity: Revisiting the Limits to Growth (or Why Industrial Society Is Inherently Unsustainable)," in D. Pimentel, L. Westra and R. Noss, (eds.), *Ecological Integrity: Integrating Environment, Conservation and Health*. Washington, DC: Island Press.
Report of the Aboriginal Justice Inquiry of Manitoba: the Justice System and Aboriginal People (1991) Province of Manitoba, Winnipeg.
Rome Declarations on World Food Security (n.d.) retrieved from <http://www.fao.org/wts/pol-icy/english/96-eng.html>.
Ross (1958).
Shue, Henry (1996) *Basic Rights: Subsistence, Affluence and American Foreign Policy*. NJ: Princeton University Press.
Sinclair, Murray, Nicholas Bala, Heino Lilles and Cindy Blackstock (2004) *Canadian Child Welfare Law*. 2nd edn., Chapter 7, "Aboriginal Child Welfare".
Soskolne and Bertollini (1999) *Ecological Integrity and Sustainable development: Cornerstones of Public Health*. Retrieved from <http://www.euro.who.int/document/ghc/Globaleco/ecorep6/pdf>.
Sterba, James (1998) *Justice Here and Now*, Combridge University Press, Cambridge, MA *U.K.* v. Iceland (1973) I.C.J. Rep. 3.
Ulanowicz, Robert (1995) "Ecosystem Integrity: A Causal Necessity," in L. Westra and J. Lemons (eds.), *Perspectives on Ecological Integrity*. Dordrecht, The Netherlands: Kluwer Academic.
Ulanowicz, Robert (2000) "Towards the measurement of Ecological Integrity," in D. Pimentel, L. Westra and R. Noss (eds.), *Ecological Integrity: Integrating Environment, Conservation and Health*. Washington, DC: Island Press.
Universal Declaration of Human Rights, pmbl. (1948) G.A. Res. 217, U.N.GAOR, 3d Sess., at 71, U.N.Doc. A/810.
Velasquez, Manuel (2000) *Business Ethics: Concepts and Cases*.
Visser 't Hooft, Hendrik Ph. (1999) *Justice to Future Generations and the Environment*. Dordrecht, The Netherlands: Kluwer Academic 6-8 Publishers.
Wackernagel, Mathias and William Rees (1996) *Our Ecological Footprint*. Gabriola Island, B.C.: New Society Publishers.
Weeramantry [Gabcikovo-Nagymaros Case (1997), I.C.J. Rep. 7 – Separate Opinion of Judge C. Weeramantry] 114 Westra, Laura (2006), *Environmental Justice and the Rights of Unborn and Future Generations*, Earthscan, London, UK.
Westra, Laura, (2004a) 'Environmental Rights and Human Rights: The Final Enclosure Movement', in *Global Governance and the Quest for Justice*, Vo1.4, Human Rights, Roger Brownsword, ed.
Westra, Laura, (2001) *Faces of Environmental Racism*, L. Westra and E. Lawson eds, Rowman Littlefield, Danham, MD Westra, Laura (1998) *Living in Integrity: A Global Ethic to Restore a Fragmented Earth*, Rowman Littlefield, Lanham, MD
Westra, Laura, (1994) *The Principle of Integrity*, Rowman Littlefield, Lanham, MD
WHO (2002)

Wood, Paul (1996) 'Intergenerational Equity and Climate Change', 8 Geo.Int'l Envt'l L.Rev. 293-307
World Bank (1999)

10

Climate Change and Poverty: Confronting Our Moral and Ethical Commitments: Some Reflections

Alicia Villamizar

Life without poverty presupposes balanced nutrition, good health, and equal opportunities in order to get basic education and adequate jobs, and all this within a safe and secure environment. The *Universal Declaration of Human Rights*, adopted by the United Nations General Assembly sixty years ago, proclaimed these among fundamental human rights for all. In the twenty-first century, these basic rights remain to be secured, as subsequently reaffirmed by the Millennium Development Goals, which are aimed at reaching a fair level of global sustainable conditions for human life. The UN Development Program's Human Development Report 2007-2008 warns: "The global warming is forcing the world towards a tipping point that could lock the world's poorest countries and their poorest citizens, leaving hundreds of millions facing malnutrition, water scarcity, ecological threats, and a loss of livelihoods." It is in this context that this paper explores the environmental implications of the current climate change negotiations.

Introduction

The Human Development Report of the United Nations Development Program 2007-2008 has warned that "the world should focus on the development impact of climate change that could bring unprecedented reversals in poverty reduction, nutrition, health and education. Global warming is forcing the world towards a tipping point that could devastate the world's poorest countries and their poorest citizens, leaving hundreds of millions facing malnutrition, water scarcity, ecological threats, and a loss of livelihoods" (UNDP 2007a). These apocalyptic words are no exaggeration. The international community—governments and citizens alike—must consider seriously the challenges of climate change that humanity must face.

Life without poverty presupposes balanced nutrition, good health, and equal opportunities in order to get basic education and adequate jobs, all this within a safe and secure environment. Those are the basic rights for humans the Universal Declaration of Human Rights stated sixty years ago (UN General Assembly 1948). In the twenty-first century, these constitute a set of minimum conditions which give our life its meaning (Agrawala 2005). Therefore, we hope that a concrete and measurable commitment toward securing the protection of these values will emerge.

In the same vein, the Millennium Development Goals (MDG) (UN Development Program 2000) aimed at reaching a fair level of global sustainable conditions for human life. In September 2000, at the United Nations Millennium Summit, world leaders agreed to a set of time-bound and measurable goals and targets for combating poverty, hunger, disease, illiteracy, environmental degradation, and discrimination against women, and placed them at the heart of the global agenda. Words of economist Jeffrey Sachs on the occasion of the presentation of the 2005 MDG progress report reflect this fact: "We have the possibility of being witnesses to the eradication of extreme poverty in the world. I am not mentioning reduction, I speak of eradication" (UN Environment Program 2005).

Within this context, the eight Millennium Development Goals should guide negotiations for a post-2012 Kyoto Protocol era which focuses on meeting the needs of the world's poor and future generations, two constituencies with a limited voice but a powerful claim to social justice and respect for human rights.

In the Latin American context, Gabaldón (2006) states:

> The biotic community, the biosphere, does not exist as a means to be merely exploited for development. The ethic of sustainable development is the fundamental of that commitment. Therefore we are compelled to know and promote the principles that make up this ethic (Gabaldón 2006, 73).

Gabaldón's statement of an ethical doctrine is reflected also in the MDG, because they represent the international community's commitment to take new actions to combat poverty, illiteracy, hunger, lack of education, inequalities between genders, mother and child mortality, disease, and the degradation of the environment. It is thus clear that the MDG constitutes, together with international agreements on climate change, a test of the world's political will to face the new social and environmental challenges of the twenty-first century.

We must see the fight against poverty and the fight against the effects of climate change as interrelated efforts. They must reinforce each other and success must be achieved on both fronts simultaneously. Success will have to involve a great deal of adaptation, because climate change is still going to adversely affect the poorest countries even if serious efforts to reduce emissions start immediately. Countries will need to develop their own adaptation plans, but the international community will need to assist them.

The Moral Environmental Commitment: What about It?

Goal number eight of the MDG—"Develop a Global Partnership for Development"—was reaffirmed by the international community at the 2002 UN Conference on Environment and Development in Johannesburg (2002) and at the 2004 World Trade Organization's meeting in Monterrey (2004). Wealthy countries are encouraged to adopt measures to ease debt and increase assistance to poorer countries and allow them access to the rich countries' markets and technology. Without the accomplishment of these objectives, it would be impossible to comply with the principles of ethical, social, and environmental morality at this critical stage of the world facing climate change.

The MDG Gap Task Force Report 2008 (UN Development Program 2008) gives us an appraisal of the progress made in reaching the MDG. The assistance of wealthy nations has decreased by 25 percent in the last fifteen years. In relation to their income, they award half of the assistance they gave in the 1960s. In 1990, the average of this assistance was at 0.33 percent of the GDP of donor countries. Today it is 0.25 percent. Only five countries—Denmark, Luxemburg, Netherlands, Norway, and Sweden—have reached the goal of 0.7 percent of their income to help poor countries. The G8 countries committed themselves to doubling their assistance to Africa for 2010. But assistance to Sub-Saharan Africa, besides the forgiveness of debt to Nigeria, has only increased by two percent between 2005 and 2006. In order to possibly comply with Goal Eight, the total debt of the world's sixty-two poorest countries would have to be forgiven, but by April 2007, only twenty-two out of forty of the poorest countries had their debts forgiven.

Under the commitment towards the MDG, developing countries have the responsibility of undertaking political reform and strengthening their governance structures for their development. The results of these efforts are to be measured at a meeting scheduled for 2015.

In 1990, 1.25 million people worldwide lived in extreme poverty, with an income of less than one dollar a day. In 2004, extreme poverty was reduced by 19 percent. This decrease had a positive effect in regions such as Sub-Saharan Africa, Central Asia, Latin America, and the Caribbean region (LAC). The LAC region has suffered from low performance, as the absolute number of poor people has increased due to population growth (Sims & Reid 2006).

These figures summarize the situation in the past decade and do not reflect the recent serious setbacks of the first decade of the twenty-first century, which have affected the growth and reduction of poverty in the region. To illustrate, if poverty is defined as the percentage of the population that earns less than 1.25 dollars per day, the Latin American region enters the twenty-first century with almost one third of its population—180 million people—living in poverty (Solimano & Soto 2006, 11-45).

It should come as no surprise that if actions to fight poverty are not urgently strengthened and enhanced, it is 800 million people who will live with less than 1.08 dollars per day in the year 2015, and not 420 million, the number set by the MDG (UN Millennium Ecosystem Assessment 2005). It is quite clear that not complying with the first goal of the MDG—that is, reducing by half the percentage of people whose income is less than 1 dollar per day and of those who suffer from hunger by 2015—makes it impossible to comply with the rest of the goals, as well as the real possibility of facing obstacles which may surface under the negative effects of global climate change. Undoubtedly, poverty in the twenty-first century has had a devastating impact on the wellbeing of so many around the world.

The frequency, intensity, duration, and magnitude of recent climatic events are forcing us to focus on global climate change, to foresee and prepare for emergencies, and to make costly and difficult decisions.

If we are morally and ethically committed to the MDG and to accomplishing the objectives set by the Convention on Climate Change (UN Framework Convention on Climate Change 1992), we should consider climate change as a key variable within the context of public and private policies.

At this point, I would like to quote a statement made by Freddy Guevara, one of our country's young student leaders, on August 2007 in the Venezuelan newspaper *El Nacional*, in regards to the serious social and political crises we are facing: "We must be awake to receive the future."

The preparations required for us to be awake and ready demands that we comprehend all adverse impacts, including the rise in the number of climate migrants, which is predicted to be 200 million by 2050 (Brown 2008, 11). It also demands that we acknowledge and appreciate what needs to be done by the developed and the developing countries to meet the challenge of climate change. This also requires an ethical and moral awakening.

Poverty and Climate Change: The Compromise

The changes experienced by the climate system are seriously affecting the ability of governments and the global society to handle the effects of climate change on their economies (Arrow 2007), natural resources (Bidegain & Camilloni 2004), and human populations (UN Development Report 2007). Our moral and ethical standards need a wake-up call—a moral shake-up—without which we will face an uncertain and bleak future.

The 1992 United Nations Framework Convention on Climate Change (UNFCCC) recognized the vulnerability of developing countries to effects of global change as well as their right to economic development. Likewise, it contemplated the obligation of developed countries to contribute to a greater extent to carry out efforts to reduce the causes of climate change, to strengthen the fight against its adverse effects, and to support economic and social development and the eradication of poverty, especially in the poorer countries (UNFCCC 1992).

Sixteen years ago, at the Earth Summit in Rio de Janeiro, the international community stated its moral commitment in regards to climate change by adopting the central objective of the UNFCCC: the stabilization of the concentrations of greenhouse gases (GHG) in the atmosphere at a level that prevents dangerous anthropogenic interferences in the climatic system.

The international community has renewed its environmental moral commitment by undertaking a negotiation process pursuant to the roadmap agreed at the Thirteenth Session of the Conference of the Parties (COP 13) of the UNFCCC (UNDP 2007b) in December 2007, in Bali, Indonesia. Since then, further work has been undertaken at negotiations in meetings in 2008 in Bangkok (February), Bonn (June), Accra (August), and Poznan, Poland (December). The Kyoto Protocol's Adaptation Fund, "a legal entity granting direct access to developing countries," is an important step forward. These negotiations are set to be concluded by the end of 2009 at the Climate Change Conference in Copenhagen, where it is expected that a new post-Kyoto agreement will be reached, as its first period ends in December 2012.

After two decades dedicated to the study of climate change and its implications for life on our planet, the evidence recently presented by the scientific community reveals a greater threat for those who have less ability to react and to adapt (IPCC 2007), which regrettably confirms the forecast of the Earth Summit and shows the need to take urgent action addressing this threat.

Towards the end of 2007, the public presentation of the UN Intergovernmental Panel on Climate Change's Fourth Report on Climate Change (IPCC 2007), and the roadmap set at the Bali Conference on Climate Change, stand as landmarks in international policy. Both events served to place the topic of climate change within the global consciousness—within the global morality—as never before. The situation is getting worse and we need to act now. Progress was made on a number of important ongoing issues for developing countries, including adaptation, finance, technology, reducing emissions from deforestation and forest degradation, and disaster management. Now the need is to implement the "common but differentiated responsibilities" principle to meet the challenge of climate change.

Are We Learning to Identify the Signal without Underestimating It?

The search has been ongoing to reach a better understanding of the complex phenomenon of climate change (Broome 2006, Raworth 2007). The global community has now acknowledged the seriousness of the problem and is becoming aware that the current environmental and social standards will not be sufficient to solve the negative effects of climate change. There has been some moral self-questioning because of the catastrophic consequences that climate change is having on global development. The need is for a new environmental ethic (Roberts & Parks 2007).

Awareness of the potential adverse impact of climate change is evident in both developed and developing countries. Even in the poorest countries, individuals and civil organizations are demanding from governments not only timely answers on environmental disasters, but also preventive and manageable measures that can reduce their negative effects (IDRS 2006). This awareness is likely to play a crucial role in securing the political commitment necessary to reach agreement on countries' obligations in Copenhagen in December 2009.

Poverty and Climate Change: A Dark Combination

The environmental morality related to climate change that has prevailed up to the present moment has helped to justify costly, and in many cases unnecessary, investments in environmental projects that have failed to counter effects of climate change. Budgets for environmental hazards related to climate are over billions of dollars per year in developed countries (Bouwer & Aerts 2006), but do not offer a guarantee of effective protection in the face of future climatic events. For example, Great Britain allocates $26 million per week for flood protection. In contrast, the cost benefit analysis related to suitable measures for climate change at a global scale shows that for every one dollar invested in adapting measures against climate change, two to three times this amount is spent on property protection. However, during a flood this enormous investment can be lost, adding up to high costs in damages associated with nutrition, health, and education (Bouwer & Aerts 2006).

These substantial investments in developed countries are accompanied by a reduction in Great Britain's financial contributions for social projects in poor countries called for under the agreements established by the eighth goal of the MDG. The IPCC (2007) considered that among the fundamental steps for a successful adaptation to climate change, developed countries needed to contribute $86 billion per year or 0.2 percent of their combined GDP to specifically improve infrastructure and build resilience for the poor against climate change in developing countries.

A preliminary estimate of the financial cost of Hurricane Gustav, which hit the Caribbean and the United States during August 2008, is nearly US$30 billion for the US alone (US Federal Emergency Management Agency 2008). It should be noted that up to September 15, 2008, five meteorological events of disastrous magnitude (Fay, Gustav, Hanna, Ike, and Josephine) occurred in the Caribbean and the Gulf of Mexico, and the total final financial cost could reach unprecedented levels (NOAA 2008). We must also remember that the worst damage was inflicted on Haiti and Cuba, two small island states among the poorest in the American continent. The losses in housing, agriculture, and electric infrastructure caused by Gustav and Ike, which hit Cuba in less than ten days between August and September 2008, add up to three billion dollars, according to unofficial estimates carried out by the Cuban government. Haiti, the poorest country in the western hemisphere, faces a chronic crisis as a result of existing poverty, which has been made worse by recent worldwide increases in food prices. It should be noted that Haiti's critical poverty was aggravated by the cumulative strikes of an unusual series of four tropical meteorological events, Fay, Hanna, Gustav, and Ike, during the last hurricane season. The case of Haiti exemplifies the nature and scope of the problem.

A dark combination of poverty and climate change poses a formidable challenge to reaching the MDG and harshly exposes the weakness of the environmental morality of the planet. More than two billion people in developing countries depend on agriculture to satisfy their basic needs and more than half of that population lives in low coastal areas, the ones more directly exposed to the effects of the rise of sea level, one of the consequences of global warming (Parry et al. 2004).

Drought and extreme rain in South America at the beginning of 2008 caused a sharp decline in agriculture in Chile and Bolivia. Powerful hurricanes hit Myanmar in May of 2008 and the Caribbean between August and September of 2008. All of these cases and many other similar examples demand a commitment of shared responsibility, which all countries on the planet have in the face of climate change, and moreover demonstrate the need for a quick and effective response to the needs of the most vulnerable (Raworth 2007, Cafiero & Vakis 2006). To reiterate, the case of Haiti shows our inability to solve the problem of poverty successfully and demonstrates the impact of climate change on poorer countries. It reminds us that the challenge of climate change calls for us to remain awake, alert, and respectful of the warnings that scientists have been providing for decades.

A recent report from the UN Food and Agriculture Organization (FAO) office in Haiti says it is important that ongoing activities, interrupted by the recent hurricane disasters, continue as soon as possible to prevent further food shortages. Prices have sharply increased and people are faced with hunger due to their homes and crops being lost. Recovery will be especially difficult for small farmers and it will have important consequences for the availability of food for a population already suffering from general chronic poverty. More than 50,000 families lost access to their usual food supply and to their means of subsistence (FAO 2008). It will not be difficult to imagine that hunger and malnutrition will get worse if the displaced rural families do not have the means to return to their land.

Only a few months before the four deadly hurricanes, the FAO had launched an emergency intervention in Haiti to promote production and improve the availability of food at the local level. However, the scarcity of food worldwide during the second trimester of 2008, which affected international markets, decreased the contributions of humanitarian assistance by developed countries to other poor areas of the world. The MDG Gap Task Force Report 2008 finds significant progress in providing debt relief to the world's poorest countries, but not in fulfilling trade and development aid commitments (UN Development Program 2008). Donors will need to increase their development assistance by $18 billion a year between now and 2010 if they are to meet their previously agreed pledges.

Conclusion

Ignoring or minimizing the risks derived from climate change on poor, vulnerable human populations can have a boomerang effect on the stronger economies of the planet (Barker 2008, Mechler et al. 2006, Stern 2007). The growing displacement of millions of people in Southeast Asia and the exhaustion of agricultural economies in the LAC region, forced by extreme events related to rain and drought experienced only in the last five years, could cause a humanitarian catastrophe in these countries (International Organization of Migration 2008; Carter et al. 2007; Bidegain & Camilloni 2004). Along with political and economic safety, environmental safety is equally important, and it could be jeopardized by hurricanes, floods, or extreme droughts. The developed world must provide assistance to avert such catastrophes.

What is needed is a new and strong ethical and moral commitment to address the issue of eradicating poverty as envisioned in the first of the Millennium Development Goals. This will enable us to adapt to the growing impact of climate change at a lower environmental and social cost.

References

Agrawala, Shardul (2005) *Bridge over Troubled Waters: Linking Climate Change and Development*. Paris: Organization for Economic Co-Operation and Development (OECD). Retrieved from <http://www.oecd.org/document/18/0,3343,en_2649_34361_36172306_1_1_1_1,00.html>.

Arrow, Kenneth (2007) "Global Climate Change: A Challenge to Policy." *Economists' Voice*, 4, no. 3. Retrieved on 10 October 2008 from <http://www.bepress.com/ev/vol4/iss3/art2/>.

Barker, Terry (2008) "The Costs of Avoiding Dangerous Climate Change: Estimates Derived from a Meta-Analysis of the Literature." *Tyndall Working Paper*, no. 17, (June) Cambridge: Pembroke College.

Bidegain, Mario and Camilloni Ines (2004) "Climate change scenarios for southeastern South America." Paper presented at the Second AIACC Regional Workshop for Latin America and the Caribbean, August 24-27, in Buenos Aires, Argentina.

Bouwer, Laurens M. and Jeroen C. Aerts (2006) "Financing Climate Change Adaptation." *Disasters*, 30(1), 49-63.

Broome, John (2006) "Valuing Policies in Response to Climate Change: Some Ethical Issues," in *The Stern Review on the Economics of Climate Change*. Cambridge: University of Cambridge Press.

Brown, Oli (2008) *Migration and Climate Change*. Ilse Pinto-Dobernig (ed.), Geneva: International Migration Organization.

Cafiero, Carlo and Renos Vakis (2006) "Risk and Vulnerability Considerations in Poverty Analysis: Recent Advances and Future Directions." *Social Protection Discussion Paper No. 0610*. Washington, DC: World Bank.

Carter, Michael R., Peter D. Little, Tewodaj Mogues and Workneh Negatu (2007) "Poverty Traps and Natural Disasters in Ethiopia and Honduras." *World Development*, 35(5), pp. 835-856.

Federal Emergency Management Agency, 2008. Retrieved on 18 October 2008 from <http://www.fema.gov/>.

Gabaldón, Arnoldo J. (2006) *Desarrollo Sustentable: La salida de América Latina*. Caracas: Grijalbo Editorial.

IPCC (2007) *Climate Change 2007: Synthesis Report. Contribution of Working Groups I, II and III to the Fourth Assessment Report of the Intergovernmental Panel on Climate Change* [Core Writing Team, Pachauri, R.K and Reisinger, A. (eds.)]. Geneva: IPCC.

Mechler, Reinhard, Joanne Linnerooth-Bayer and David Peppiatt (2006) "Disaster Insurance for the Poor? A Review of Micro-Insurance for Natural Disaster Risks in Developing Countries," in *Provention/IIASA Study. Provention Consortium*. Ginebra,
France, Kenya,UK, USA, Netherlands, Malaysia.

National Oceanic and Atmospheric Administration (2008) *Hazards/Climate Extremes*. National Climatic DataCenter,U.S.Department of Commerce.

Operation Green Leaves (2008) retrieved on 20 September 2008 from <http://www.blogtalkradio.com/OGLHaiti/blog/2008/09/13/Haiti-Calling-Post-IKE-Update-on-Environmental-Disaster-Caused-by-Deforestation>.

Parry, Martin L., Cynthia Rosenzweig, Ana Iglesias, Matt Livermore and Guenther Fischer (2004) "Effects of climate change on global food production under SRES emissions and socio-economic scenarios." *Global Environ. Chang.*, 14, 53-67.

Raworth, Kate (2007) "Adapting to Climate Change. What's Needed in Poor Countries and Who Should Pay." Oxfam Briefing Paper No.104. Oxford: Oxfam International.

Roberts, Timmons and Bradley C. Parks (2007) *A Climate of Injustice: Global Inequality, North-South Politics and Climate Policy.* Cambridge, MA: MIT Press.

Sims, Andrew and Hannah Reid (2006) *Up in Smoke? Latin America and the Caribbean: The Threat from Climate Change to the Environment and Human Development.* Third Report from the Working Group on Climate Change and Development, New Economic Foundation.

Solimano, Andres and Raimundo Soto (2006) "Economic Growth in Latin America in the Late Twentieth Century: Evidence and Interpretation," in Andrés Solimano (ed.), *Vanishing Growth in Latin America: The Late Twentieth Century Experience*. Chile: Editorial Edward Elgar.

Stern, Nicholas (2007) *The Stern Review: The Economics of Climate Change.* Cambridge: University of Cambridge Press.

United Nations Development Program (2008) *Millennium Development Goal 8. Delivering on the Global Partnership for Achieving the Millennium Development*. MDG Gap Task Force Report 2008.

United Nations Development Program (2007a) *COP 13th Conference on the Parties of the United Nations Convention on Climate Change*. December 2007, Bali, Indonesia.

United Nations Development Program (2007b) *Human Development Report 2007/2008. Fighting climate change: Human solidarity in a divide world*. New York: Palgrave Macmillan Pub.

United Nations Development Program (2005) *Millennium Ecosystem Assessment: Living Beyond our Means: Natural Assets and Human Well-being (Statement from the Board)*.

United Nations Environment Program (2005) Communication and Public Information Unit. Daily News. Retrieved on 20 September 2008 from http://www.pnuma.org/informacion/noticias/2005-01/24ene05e.doc.

United Nations Food and Agriculture Organization (FAO) (2008) "Haiti." Retrieved on 15 October 2008 from <http://www.fao.org/isfp/country-information/haiti/it/>.

United Nations Framework Convention on Climate Change. 1992. Available at www.unfccc.de/resource/convkp.html.

United Nations General Assembly (1948) *Universal Declaration of Human Rights*. Resolution of the General Assembly 217 A (III) 10 December.

Part IV

Civil Society

11

Soft Power, NGOs, and Climate Change: The Case of The Nature Conservancy

Katrina S. Rogers

Soft power is a conceptual framework that posits that the ability to influence others through attraction rather than coercion or payment is an important form of state power in world politics. This chapter presents evidence that the rise of non-governmental organizations (NGOs) exerting substantial soft power is resulting in two new influences in international relations: creating a new, stronger multilateral platform for policy debate and action, and establishing opportunities for more informed debate and decision-making processes in the environmental arena, particularly in the area of climate change. Coined by the political scientist Joseph Nye, the term soft power refers to the attractiveness of a country's culture, political ideals, and policies, and the way in which these attributes are used. When a country's policies are seen as legitimate and attractive in the eyes of the world, soft power is enhanced. Soft power is real power in that states use it in order to achieve their objectives. Yet, it can be haphazard as well, as having cultural attributes or political ideals that are attractors is part of a changing milieu of preferences not always controlled by states. Over the last several decades, the dynamics of the international state-centric system have shifted. Once characterized by state actors and bipolar relations, international relations are now characterized by multiple actors, and an increasingly unstable unipolarity.

Examining The Nature Conservancy's (TNC) climate change program as one example reveals indications that soft power is increasingly a tool used by TNC to influence other actors. Evidence from this case study supports the notion that the current fluidity of international relations has created the conditions for resource-rich NGOs to have a significant impact on global environmental problems in general and climate change in particular. Soft power can be seen as an explanatory framework in which we can discuss the possible evolution of a more ethical system of collaboration and attraction, rather than coercion and force.

Soft Power and Climate Change

As the students blankly stared at me, I realized that most of them were not born when the Wall came down between East and West Berlin. The world in which they grew up was the dynamic and multipolar system of the new world order of the 1990s, followed by the optimism and economic prosperity that characterized the technology boom and the years of the Clinton administration. American political culture during the years of the Cold War, and the bipolarity dominating international relations then, is not only conceptually different from that of the culture of the younger generation, but also one that contrasts sharply with the complexity of state action and international behavior today.

The Cold War era was a period of relative stability in which two superpowers, the United States and the Soviet Union, played out their conflicts in other parts of the world, usually through surrogates in Asia and Africa, and throughout Latin American countries. The world was their chessboard, but Europe would be ground zero if the Cold War ever got hot. Poland and the Eastern Bloc countries looked nervously over their shoulders at the superpower to their east. While some of the historical content is yet cogent, international relations is no longer a static system of set roles and predetermined power relationships.

The Emergence of International Power Relationships

In the seventeenth century, Hugo Grotius outlined the international system as a chaotic shifting series of interrelated power relations between states (Grotius in Edwards 1981). Grotius' view of a hierarchical state-centric system provided the dominant paradigm for nearly four hundred years. Today, the advent of powerful non-state actors challenges the old organizational perspective. Supra-state alliances, multinational corporations (MNCs), stateless terrorist organizations, and non-governmental organizations of all stripes now flex their muscles in what seems to be an increasingly anarchic system, wielding influence and power in ways that could not be imagined in the seventeenth century or even fifty years ago.

Political scholars have long been intrigued by this increase in the number of non-state actors within the international system, and a good deal of scholarship has been devoted to understanding these dynamics. Stephen Krasner (1983), for example, outlined the concept of "regimes" as a way of understanding how different entities, both state and non-state, may come together to manage a particular global problem.

International regimes for oceans, space, Antarctica, and other regions seen as the global commons have emerged in recent decades. Oran Young and Marc Levy's (1999) work on international cooperation has also revealed some of the dynamics about NGOs' ability to act within the international system.

Other themes have emerged, including the increasing impact of MNCs and NGOs that have come to influence the actors and to themselves engage in multilateralism. Clark (1995) argued that environmental NGOs had developed channels of communication independent of traditional state negotiation and communication. For example, environmental NGOs can lobby independently of states within the UN system, the Organization of American States, and the Council of Europe. Environmental NGOs, she said, also have the agility and flexibility to engage with other NGOs in ways that states cannot. This collegiality has the potential to build an international civil society in their sector around their issue.

Joseph Nye: Soft and Hard Power

Joseph Nye's framework of hard and soft power is helpful in understanding how NGO behavior can be seen as a set of nuanced, coordinated actions that employ elements of soft and hard power. Hard power and soft power dynamics range from war and other acts of coercion on the "hard" end, to inducement, agenda-setting, and attraction on the other end, where soft power lies. In his conception, hard power is defined as the ability to change what others actually do through coercion. Hard power is generally quite tangible, such as military and economic power wielded more-or-less blatantly by state actors. A visible example is the US war in Iraq, where economic incentives were used to attract coalition partners as well as the clear military force wielded in a country to depose a dictator. Other examples of hard power are the economic sanctions imposed on Iraq between 1990 and 2002, or the force used by Israel to contain militant Palestinian forces (Nye 1990).

Soft power is the capacity to shape what others want through affinity or enticement. Examples of soft power could be a country's culture and its attractiveness for another country or the attractiveness of international technology to make other countries want that technology from another country. Nye argues that a country has more soft power if other actors respect and admire the values and character of the nation. He suggests that soft power is an important undercurrent in diplomacy and that it facilitates the building of alliances. For example, in an article about America's decline in soft power, he points to a Eurobarometer poll that found that a majority of Europeans believe that Washington has undermined efforts to fight global poverty, protect the environment, and keep peace in recent years. Such attitudes reduce the ability of the United States to achieve its goals without resorting to payment or coercion (Nye 2004, 16). In this way, the diminishment of soft power increases the likelihood of the use of hard power, which has both economic and military consequences for state actors and possibly others who get drawn into the melee.

Nye explicates a dynamic framework in which hard and soft power are not mutually exclusive, but rather both used by states as multivariate means of obtaining what they want. This provides a useful conceptual framework for understanding NGOs, which have often been seen solely as purveyors of soft power, using tactics such as inducement, agenda-setting, and attraction because those are the tools they have at hand. Examples include Amnesty International's Human Rights Watch reports, which attempt to raise awareness and influence domestic populations to pressure states to adhere to basic human rights. Another example is the World Business Council on Sustainable Development's endorsement of the UN Millennium Development Goals as they relate to sustainable development. Both use agenda-setting and attracting others to their organizations' mission and vision.

An important element in Nye's explanation of soft power relevant to states is that the legitimacy of state policies and the values that underlie them can attract other states to their point of view. His critique of the former Bush administration is a case in point. The questionable use of hard power in Iraq undermined US legitimacy abroad. An analog in an NGO context is the need for such organizations to attract human and financial capital to their missions through constantly framing and re-framing their work in attractive ways. A specific instance is the organization Earthjustice, with its beautiful ads in international airports with the tagline, "Because the Earth needs a good lawyer." In these few words, this NGO explains its mission and attempts to attract others, which further legitimizes its cause.

The most common critique of Nye's hard and soft power explanatory model comes from the realist perspective, which argues that hard power is the only form of power that is unequivocally understood and used successfully in international relations. The Bush administration (2000-2008) was a good example of not subscribing to the soft power form of politics. For example, former Secretary of Defense Donald Rumsfeld argued that the United States was strong enough to do what it wanted,

Figure 1
Hard and Soft Power Dynamics

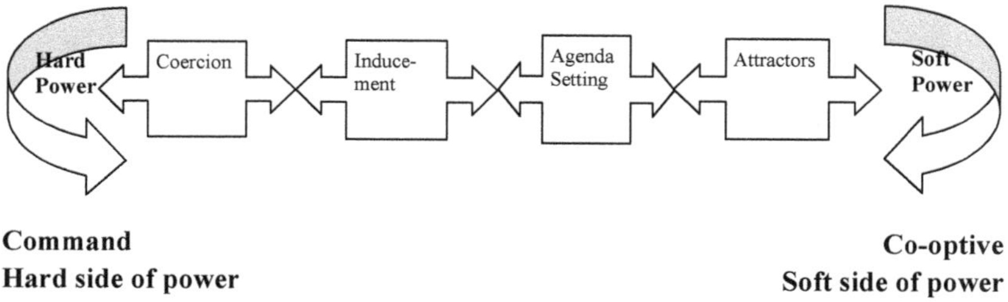

Figure 2
Earthjustice Ad as an Example of Soft Power Attraction (2010)

and there was no need to negotiate or pretend to build alliances in a world where the United States is the superpower. The former US ambassador to the United Nations, John R. Bolton, made several remarks that can be read as skeptical of multi-governmental processes in particular and soft power in general (Bohorquez 2005). In light of the US position as the world's only superpower, soft power is not necessary to achieve foreign policy goals. In a more nuanced critique of the hard-soft power model, others have suggested a conceptual confusion exists within the term, since power is always used in a situational context, which may or may not be judged as coercive depending on which actor is viewing the situation. The party under pressure may not make any distinction between hard (coercive) power and soft (non-coercive) if it becomes compelled to do something it would not otherwise have done (Ogoura 2006). Both critiques raise interesting questions for NGO behavior, since organizations tend to view most of their influence as non-coercive and thus within the range of soft power.

The Nature Conservancy

To add some clarity to understanding NGOs' use of soft power, it is useful to consider other ways in which these organizations exert their influence or otherwise induce others to support their agendas. NGOs can be transformers, converting the elements of soft power to effective hard power for the purpose of manipulating other actors. This power is often coercive, even though it does not arise from bases traditionally perceived as powerful. Powerful NGOs such as TNC have the ability to move all over the chessboard of international relations. With their cachet as an NGO and their resources, they often bypass the protocols that states must observe. This ability to be agile means that they can transform their soft power to the hard power that international institutions and states use.

The Nature Conservancy, created in 1951, is one of the largest private nonprofit conservation organizations in the world. Founded on the principle that the loss of biodiversity can be stemmed by purchasing land and protecting it, this group has been the trailblazer for organized environmental movements in the United States and throughout the world. Currently an entity with more than a million members, an annual budget of $900 million (The Nature Conservancy 2007), and a staff of 3,200 in thirty countries and all fifty US states, it has resources far beyond those of most non-governmental organizations and the conservation budgets of many countries.

TNC's organizational culture is innovative, and over the years it has developed sophisticated conservation tools in pursuit of its mission. As one example of its innovative ability, TNC was one of the first organizations to create a large carbon sequestration project in Latin America. TNC has also shown an exceptional ability to work both in the halls of governmental power and in corporate boardrooms. Staffers pride themselves on being able to move smoothly between government and industry (TNC staff 2007).

The organization's record of success and its moderate strategy indicate that it is worth taking a closer look at how it uses its ample financial resources to address a major environmental issue of our time: climate change. This research suggests that TNC has developed the ability to use transformed power. Transformed power enables TNC to utilize hard power strategies and tactics to accomplish its mission, albeit in the nominally restricted position of soft power. Understanding this nuance assists scholars in revealing the level of complexity and strategies of engagement in evaluating general NGO behavior using the specific example of TNC's climate change work to test this idea.

TNC: Mission and Accomplishments

The heart of the mission of The Nature Conservancy (2007) is the ambitiously stated goal "to preserve the plants, animals and natural communities that represent the diversity of life on Earth by protecting the lands and waters they need to survive." In the early days of its work, TNC secured places simply by purchasing them with private money or accepting gifts of land. Over time, its work has become increasingly sophisticated and often innovative, using tools such as revolving land protection funds, debt-for-nature swaps, public-private partnerships, scientific frameworks, international conservation, and policy advocacy. Many of these tools were initially invented by TNC and subsequently adopted by other organizations. As another indication of its impact, many conservation professionals started their careers at TNC and went on to lead or begin their own organizations, including the Audubon Society, the African Wildlife Foundation, Ecotrust, Conservation International, the Conservation Fund, the Grand Canyon Trust, and the Minnesota Land Trust.

As an organization, TNC's accomplishments are impressive. Since its founding, TNC has protected more than 117 million acres of land and 5,000 miles of rivers, and operates 100 marine conservation projects worldwide. The Conservancy has assisted in developing a hemispheric biological inventory to track some 50,000 species and ecological communities, and mobilized billions of dollars in public funds to acquire and protect important nature areas. This figure amounts to $29 billion in public money raised since 1986 (The Nature Conservancy 2007).

As TNC's organizational structure evolved, it became clear that in order to succeed at its mission of protecting all life on earth, it would have to make a commitment to working in other countries where biodiversity is rich and critical for the health of the planet. Species richness increases from the poles toward the equator, and a similar pattern is reflected in the marine environment (Bryant 2006). It is no coincidence that TNC, Conservation International, and the World Wildlife Fund have all turned their attention to Latin America and the Asia-Pacific region. Considering the difficulty in countries like those in the former Soviet Union and the Eastern European region, it is a logical choice to focus conservation attention on the areas of the world where biological diversity is at its highest and governments are often more responsive and possible to work with over the long term. Although the majority of the species on earth exist between the Tropics of Cancer and Capricorn, TNC's entry into the issues of climate change has expanded the organization's concerns all the way to the North and South Poles.

Soft Power in Action

Soft power can be described as a kind of "tactical jiu jitsu," where an NGO leverages its influence to have governments use their hard power resources to accomplish the NGO's mission. A good example is The Nature Conservancy's well-known Parks in Peril program. Part of that program was to get governments to create national parks and address poverty that created environmental degradation. TNC spent a great deal of time and resources trying to convince Latin American governments that setting aside land was part of becoming modern. This is a clear case of Nye's broader point that soft power can also include the allure of certain ideas and cultural norms. Another, perhaps more coercive, example was when TNC successfully lobbied the US government to delay loans to Latin American countries because environmental safeguards were not in place. By affecting those decisions, TNC was transforming its soft power (attraction and inducement) to harder power (economic sanctions).

In an article about human rights and environmental NGOs, Clark (1995) argued that environmental NGOs had developed channels of communication independent of traditional state negotiation and communication. In elaborating on this discussion of the effectiveness of environmental NGOs on the international stage, I can add one from TNC's experience: the power of science or the Big Idea to move and influence other actors' agendas. TNC discovered that it could piggyback on developing countries' interest in science and development along with their desire to be considered scientific and modern. This push toward the hegemony of science worked in TNC's favor, as it was able to show and convince leaders and other policymakers that a key element of a developed society was a conservation program that set aside and protected critical habitats. This dynamic came into play particularly when working with the top leadership of a country, including agency heads, ministers, and even presidents.

An example of this dynamic related by a TNC senior staffer occurred during the 1980s and 1990s as American environmental NGOs were jockeying for position in Latin America and other places. Predictably, competition tended to focus on the upper levels of management. While the scientists and conservation staff employed by all may have cooperated in terms of sharing information and knowledge, senior leaders were positioning their organizations to be attractive to funders. The cooperation that was evident tended to be either upward, that is, with government leaders, both funders and grantees, or downward to on-the-ground partners in various countries.

Peer-to-peer cooperation was much less evident. "TNC, Conservation International, and the World Wildlife Fund would never have met together back then," confided one staffer (TNC staff 2007). Collaboration was driven only by money, if the funders demanded it. A senior staffer reported that a Latin American colleague at a conference of Latin American NGOs said they did not want collaboration among the big American NGOs. "The more you compete with each other, the more we can get from each of you," was the revealing admission (TNC staff 2007). This is probably not a startling revelation, but it clearly shows that the recipient states' organizations understood their advantage even as they solicited donors. They quickly learned how to use the competition among the big NGOs to glean more funds from each of them as they competed with each other.

TNC and the Earth Summit

The United Nations Conference on Environment and Development (UNCED) was held in Rio de Janeiro in June 1992. Informally called the Earth Summit, the conference comprised 172 countries and 2,400 representatives of NGOs meeting for two weeks to highlight and discuss contemporary environmental problems and issues. Non-governmental organizations, however, met in a separate informal conference across town from the location of the formal talks. Seventeen thousand people attended the NGO conference, from private individuals to heads of state.

The Earth Summit emerged from global conversations about the problems of economic development in the face of environmental degradation occurring planetwide. This UN conference attracted people from all over the world and resulted in increased global awareness of the environmental problems facing human societies. Now, more than a decade later, the summit has been criticized for not yielding much beyond a watered-down Agenda 21, which is a framework for sustainable development. It did make a clear statement that humanity needs to transform its attitudes in order to change priorities and take action if conditions for both humans and the natural world are to be improved. The complexities of the problems facing all of us, but particularly states,

NGOs, MNCs, and others, were drawn in stark terms at the Earth Summit. It was clear that poverty and excessive consumption are both having an extraordinary and deleterious impact on the environment.

For the first time in history, a concerted worldwide plenum convened to articulate that humanity has the opportunity, ability, and mandate to address the warming of the earth's atmosphere. The three warmest years on record have all occurred since 1988, and nineteen of the warmest twenty years have occurred since 1980. Climatologists reporting for the UN Intergovernmental Panel on Climate Change (IPCC) indicate that the world is experiencing global warming caused by human activities. Global average atmospheric CO2 rose from 280 ppm at the start of the Industrial Revolution (approximately 1750) to 381 ppm in 2006. The present concentration is the highest level during the last 650,000 years and probably during the last 20 million years. The growth rate of global average atmospheric CO2 for 2000-2006 was 1.93 ppm y^{-1} (or 4.1 PgC y^{-1}), and it was 1.5 ppm y^{-1} for the previous thirty years (http://www.globalcarbonproject.org/). If current trends continue, atmospheric CO2 concentrations are expected to double from pre-industrial levels during this century. That will probably be enough to raise global temperatures from 3 to 10 degrees Fahrenheit. As the temperature rises, a myriad of effects will be felt that have implications around the globe: the ocean's rise will lead to the inundation of low-lying areas and islands, crop yields will become more vulnerable, ecosystems will shift, storms will become more frequent and more violent, plants and animals will die or be pushed to ever-narrowing ranges, and humans may find themselves forced to move from place to place to escape an increasingly unstable climate.

At the Earth Summit, TNC was present in its role as a large US-based conservation organization doing work in Latin America (TNC staff 2007). The year 1992 was the height of the rainforest protection frenzy (TNC staff 2007). Through the mid-1980s and until the early 1990s, virtually every week an article appeared in a major American newspaper about the destruction of the rainforest. Major fires were being set in the Amazon, and the American public became aware that the destruction of these forests was part of a continuing global environmental degradation. TNC was at the Earth Summit, along with the World Wildlife Fund and Conservation International, the other major players in the region, in order to generate media activity and highlight its conservation efforts and philosophy.

In 1992, the World Business Council on Sustainable Development was created, as well as the World Bank's Global Environmental Facility, a fund for environmental protection. The Summit was attended by the president of TNC, John Sawhill, who took the opportunity to increase TNC's visibility in international conservation issues. During the meeting, he met with the president of Brazil as a symbolic and real gesture of the importance of conservation in that country's agenda. This meeting made the front page of the newspaper in Rio de Janeiro, the *Jornal do Brasil*. According to one staffer, the meeting was a kind of "eco-theater," in which the point was to create an impression of cooperation that could be parlayed into real cooperation and conservation (TNC staff 2007). TNC at that time was trying to establish itself as a global player. Even though there was not much opportunity for substantive conversation, the "eco-theater" turned out to be critical to TNC's mission.

TNC's presence at the summit demonstrated the soft power that the organization commands in its issue areas. The "eco-theater" allowed them to meet with ministers, top officials, and presidents of other countries, which in Latin America had a significant impact on managers throughout the government. TNC understood that, particularly in Latin American political culture, when the president showed interest in conservation by public proclamations, it was seen as important to local government officials.

TNC's participation in the Rio Summit helps to illustrate Clark's point earlier in this essay that NGOs can and do harness additional communication channels. TNC was able to jump diplomatic protocols. A large American NGO working in Latin America automatically brings with it the cachet of international influence. Local partners want to team up with a bigger organization that can provide access to funds and networks of influence for their own issues and concerns. At this particular time, TNC was trying to get its Parks in Peril program, one of its most successful programs and one that still continues today, going. It was a highly competitive setting, with NGOs competing for money from North American private funders, government agencies, and foundations. Money flowed to TNC from these sources as a result of media attention combined with TNC's success in obtaining the support of leadership in Latin American countries.

Soft Power and Diplomacy

A senior TNC staff member shared with me an event that happened in 1989, when a formal state visit was scheduled between US Vice President Dan Quayle and the president of Bolivia, Jaime Paz Zamora. When TNC learned of the planned meeting, it decided that it might be able to influence this meeting in a way to insert conservation issues into the formal discussion. TNC called the planner for the president of Bolivia and asked, "What would you think if Vice President Quayle brought up conservation in the conversation?" The answer was that the Bolivian president would welcome the discussion (TNC staff 2007). (What else could he say?) Then TNC called Vice President Quayle's office and said that the president of Bolivia would be amenable to including conservation on the agenda. TNC subsequently drafted talking points for both offices. This meeting helped TNC forward its agenda of promoting several important projects in Bolivia at the time, including the Noel Kempff National Park expansion, which quadrupled its size to two million acres. States could never do what TNC had just orchestrated, because TNC can bypass protocols, whereas states are much more restricted. NGOs are not stuck in the same hierarchy of power relations, but rather move at multiple levels and use soft power to achieve hard goals. This example both contributes to Clark's point about the agility and flexibility in diplomacy that is less formalized than state to state relations, and gives another indication of TNC's use of soft power to orchestrate actors with hard power capacity, in this case, the United States and Bolivia.

TNC's Climate Change Initiative

The roots of TNC's climate change policy go back nearly two decades to the creation of its first carbon sequestration project to protect the endangered Latin American rainforest in 1990. The 150,000-acre Mbaracayu Preserve Project in Paraguay was purchased using funds contributed by a US electrical power producer. This purchase prevented the clear-cutting for which the forest would be slated under normal circumstances. The AES Corporation considered the $2 million donation to protect the rainforest as a way to offset carbon emissions from a new coal-fired power plant being constructed in Hawaii (TNC staff 2007).

More recently, TNC has begun to commit considerable attention and resources to the issues of climate change, giving us an opportunity to examine how this work, although somewhat different from its initial work in Latin America, is informed by those strategies and tactics while utilizing many of the same channels.

In December 1997, the 3rd Conference of the Parties to the Treaty on the UN Framework Convention on Climate Change met in Kyoto, Japan, an event which resulted in what is popularly known as the Kyoto Protocol.

The first steps were taken in agreeing to ameliorate climate change (UNCCC, 2008). The agreements, which came into force on February 16, 2005, have been criticized for not being stringent enough and for not including the biggest emitters of fossil fuel emissions. Saving natural communities now includes preserving land to provide migratory routes for imperiled species. These routes are increasingly important in the face of changing climates. To this end, political advocacy and science-based argumentation have become tools to bolster TNCs conservation mission. Only countries are signatories, but many multinational corporations have pledged to expedite targets. However, another singular weakness is that the Protocol expires in 2012.

Several conferences and meetings have been held to begin extending the Protocol and to determine how to bring in the three most important state actors who are currently significant outliers: the United States, China, and India. Talks have occurred in various parts of the world, most notably in Poznan, Poland on December 2008, where a vision was hammered out. The talks culminated in high level negotiations in December 2009 in Copenhagen. The Web site sponsored by the UN Framework Convention for Climate Change (http://unfccc.int/2860.php) gives the most up-to-date information on international negotiations.

The United States was present at Rio and is a signatory; however, the US Senate has refused to submit the treaty for ratification, and former President Bush (2000-2008) stated that, for numerous political, economic, and science-based reasons, he would not sign the treaty into law even if the legislation emerged from the Senate. President Obama's administration is shaping up to take a more proactive stance on international climate change frameworks. He sent a team to Poznan, Poland, and has remarked that his administration will take an aggressive position on climate change, calling it a "matter of urgency and of national security and has to be dealt with in a serious way" (Reuters, December 9, 2008, http://www.reuters.com/article/newsOne/idUSTRE4B86R920081210).

Interestingly, and despite the political and economic unpopularity of the Kyoto Protocol, many US states, companies, and other actors have taken action to implement some or all of its standards. Not surprisingly, TNC has been at the forefront of organizations filling the gap left by the US government's inaction. First and earliest was its own Climate Change Initiative. Second, it has supported legislation in the US House and Senate.

Flowing directly from its mission to protect the diversity of life on earth, the major goals of TNC's Climate Change Initiative are as follows:

- To pursue a comprehensive national policy to reduce carbon emissions;
- To support and advocate for policies at state and local levels within the United States (such as the more stringent regulations of California and cities like Seattle, WA) to reduce carbon emissions;
- To assess the ecological risks for plants and animals, most of which will be impacted by climate change;
- To mitigate impacts of climate change for species where possible by, for example, protecting ecosystems large enough to allow for species migration; (Saving natural communities now includes saving land for migratory routes for imperiled species. These routes are increasingly important in the face of changing climates. To this end, political advocacy and science-based argumentation have become tools to bolster its conservation mission.)
- To demonstrate the importance of action by saving land and protecting biodiversity;
- To create specific climate action projects that seek to address global deforestation and reduce carbon emissions (TNC 2008).

TNC and the Climate Stewardship Act

Political activism now includes advocacy for a comprehensive federal policy aimed at reducing carbon emissions. Since the United States is the biggest contributor to global warming, TNC's efforts are devoted to educating and lobbying the federal government to institute policy changes. The tip of the spear in this strategy has been to build broad support in Congress and among key constituencies for the Climate Stewardship Act, a piece of legislation narrowly defeated in October 2003. Sponsored by Senators John McCain and Joe Lieberman, this act embodies principles that will have significant impact on activities in the United States that contribute to global warming. The act, which is summarized on TNC's Web site, has the following elements:

- All major sources of emissions that contribute to global warming would be required to limit their total emissions to year 2000 levels by 2100. This includes the energy, industrial, and transportation sectors. Utilizing a market-based approach, the act would establish a nationwide set of emission standards, and create a trading system that allows companies to buy and sell "carbon credits" (similar to the local Chicago carbon credit system). The Chicago Climate Exchange (CCX) is a voluntary trading market to help companies prepare for future US greenhouse gas caps. CCX members enter legally binding agreements to cut their emissions by a set amount and time. If they reach those targets, they generate credit they can sell or bank. If they fall behind, they must purchase credits on the Exchange. Recently, the Exchange has been working on a deal in which its members can use carbon dioxide emissions allowances from the European Union. If successful, this could be the first trading link between greenhouse gas emitters in the United States and the European Union's Emissions Trading Scheme.

- These carbon credits allow companies to either become more efficient (pollute less) or purchase carbon credits from other companies. Since the number of credits is finite, emissions are reduced.
- Other financial incentives are provided for various industries to earn additional allowances for reductions achieved if accomplished before the mandatory reduction requirement (US Congressional Legislation 2008).

The intent of the Stewardship Act is carefully worded to obtain buy-in from a Congress and a public that is equivocal about climate change initiatives, particularly in the midst of a recession. This legislation is currently in committee, and will most likely remain there until the new Congress convenes in January 2009. While limits on emissions are mandatory, there is an emphasis on market-based approaches, a trading system, and financial incentives. TNC has been criticized for taking this more moderate and, to some, weak stance. Viewed with an understanding of soft power, this stance can be seen as a way to coax the United States to flex its hard power in accomplishing TNC's goals and ultimately its mission.

Now that the US House of Representatives has adopted legislation with cap-and-trade provisions, this instance provides yet another example of TNC's use of transformed power; that is, to convert its soft power and move along the continuum toward hard power politics.

TNC and Science as a Tool

Nearly every observer of humanity, from Adam Smith to Alexis de Tocqueville to Milton Friedman, supported by a great deal of social science research, agrees that self-interest (with any of its possible qualifiers: enlightened, calculated, rational, etc.) is an important factor driving the decisions of most of the people, most of the time. However, science has become a remarkably effective tool used to sway people's understanding about what is best for them (i.e., in their own self-interest).

Now, with more than 700 staff scientists, TNC researches and presents policy options to the public and to decision makers. The organization's tactics are strategically focused and aggressive, not unlike its efforts in the 1990s to promote rainforest conservation by keeping the facts about the scientific basis for that protection squarely and repeatedly in the public's eye. Its chief climate change scientist, Earl Saxon, frequently publishes about the profound changes to be expected as a result of global warming. A 2005 posting (Saxon et al. 2005) on its website discusses the NASA study that reported that 2005 was the warmest year on record. While not directly taking on the US executive branch by criticizing climate change policy, TNC does, through its scientific research and reporting, make its strong point of view clear to all. Saxon et al. affirm that "Significant disruptions to natural ecosystems are widely expected as a result of climate change and … environmental changes will create additional stresses on those plant and animal populations whose adaptive responses are unable to keep pace" (ibid. 53).

In a radical departure from the days of the Earth Summit, TNC and Conservation International are now engaged in a Joint Initiative on Climate Change and Conservation in which their scientists collaborate and publish essays together on the future of the planet. According to one TNC staffer, this initial collaboration was driven by an outside funder but now has a "real" quality to it, at least among staff-level scientists. This collaborative thinking around science is an interesting evolution and one that reflects the necessity of bringing as many resources as possible to bear on these complex, global problems.

Using science as leverage to influence policy, TNC now advertises five examples of climate action projects. All of them meet the criteria of either helping to absorb carbon dioxide (aka carbon sequestration) or safeguarding large tracts of land. One example is the Guaraquecaba Environmental Protection Area in Brazil. Other projects are the Rio Bravo Conservation and Management Area (Belize), Noel Kempff Mercado National Park (Bolivia), Midwest Forest Restoration (Ohio and Indiana, US), and the Louisiana Bayou Pierre Floodplain (US).

In the project in Brazil, 50,000 acres of tropical forest are being restored. This restoration includes an analysis of the major threats to the forest, one of which is large-scale deforestation as a result of Asian water buffalo ranching. The project has begun to remove buffalo and to restore forests to land which had been cleared for pasture. TNC scientists believe that the protection area, once reforested, will help to absorb carbon dioxide and become a mitigating factor in climate change. A key element of this project is the number of additional partners who are participating, including American Electric Power, General Motors, Texaco, and Sociedade de Pasquisa em Vida Selvagem e Educação Ambiental (SPVS). In addition to ecological restoration and carbon sequestration, TNC touts the project in terms of its longevity (perpetuity) and its positive impacts on climate change in that, as the planet continues to warm, the project will provide plants and animals with migratory routes and survival islands.

The back story to the Brazilian project is the use of science to forge the partnerships that enabled this project to be created. It is clear that TNC sees science as an important tool in its soft power toolbox. It would also appear that one of its weaknesses is that little exists in its toolbox about human behavior and social change. On the surface, TNC has far more physical scientists than social scientists among its staff. One possible evolution of TNC may be, in the coming years, an acknowledgment that environmental problems are fundamentally human problems, forcing TNC into harder positions of advocacy and political activism.

Conclusion

Using Nye's soft-hard power model, the ways in which power is used by NGOs to influence actors, including states, in international relations become apparent. The current fluidity of international relations, the growing urgency of the climate change crisis, and the example of TNC has created conditions for the more resource-rich NGOs to have a significant impact on global environmental problems.

Nye's model offers a way to deepen scholarly knowledge and to examine and better understand the complex ways in which actors engage in their work. With TNC as an example, the soft-hard power model works to illustrate the nuanced ways that TNC operates as a resource-rich and environmentally focused NGO. In the case of TNC, it would appear that there are many times when its soft power is coercive, such as in the case of pressuring states to implement sanctions against other states to promote good conservation practices such as land protection or to stop poor practices such as clear-cutting. To the actor being pressured, such as a Latin American country, TNC's use of power, albeit indirect, may still seem coercive. The framework is useful as a way to understand the complex, multivariate ways that actors engage in international work, and to reveal explanations about NGO actions and use of power.

Equally important is to recognize that new strategies and tactics have been added to the toolbox of NGOs. These new tools include the power of political advocacy and the persuasiveness of science in creating positive change. With advocacy, TNC walks a careful line of moderation. With science, it has become more assertive in promoting science-based projects.

Over the past fifty years, The Nature Conservancy has transformed the traditional soft power of a nonprofit, non-governmental organiza-

Figure 3
Significance of TNC Influence in Climate Change

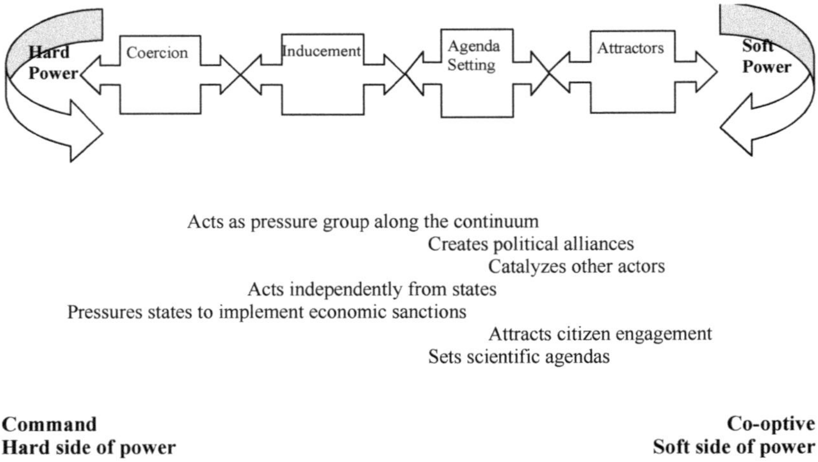

tion into a new and effective form, not directly coercive as states' hard power, but a transformed power giving additional leverage as well as gaining independence from established norms and protocols. As the effects of climate change inflict ever more dramatic and predictably more devastating effects on the earth's ecosystems, TNC will be compelled to speak out even more aggressively if it is to continue to succeed on an international scale. Because climate change will not reverse course any time soon, TNC will find it difficult, if not impossible, to also reverse course.

Although TNC's mission "to preserve plants, animals and natural communities" remains unchanged, over the last half-century its understanding of what that mission implies has expanded considerably. The Nature Conservancy's work demonstrates that policy formation is about human behavior, and ultimately climate change is about humanity: its role in the evolving crisis, and its adaptive and survival responses. TNC must add the human element to the implementation of its mission. The addition of a strong base in the social sciences will be critical to meaningful work in the future. Examining The Nature Conservancy's climate change program as one example reveals indications that it is increasingly using soft power to influence other actors. Evidence from this case study supports the notion that the current fluidity of international relations has created the conditions for resource-rich NGOs to have significant impact on global environmental problems in general and on climate change in particular. Soft power can be seen as an explanatory framework for discussing the possible evolution of a more ethical system of collaboration and attraction, rather than coercion and force.

References

Bohorquez, Tysha. 2005. Soft power – The means to success in world politics. *UCLA International Institute*. http://www.international.ucla.edu/article.asp?parentid=34734.

Bruner, J. 1985. *Acts of meaning*. Cambridge, MA: Harvard University Press.

Bryant, Peter J. 2006. Biodiversity and conservation. University of California Irvine. http://darwin.bio.uci.edu.

Clark, Ann Marie. 1995. NGOs and their influence on international society." *International Affairs* 48.

Edwards, Charles S., and Hugo Grotius. 1981. *The miracle of Holland*. Chicago: Nelson Hall Press.

Keohane, Robert, and Joseph Nye. Power and independence in the information age. *Foreign Affairs* 77, no. 5: 81-94.

Krasner, Stephen (ed.). 1983. *International regimes*. Ithaca: Cornell University Press.

Lumsdaine, David Halloran. 1993. *Moral vision in international politics*. Princeton: Princeton University Press.

Nye, Joseph, I. 2005. *Soft Power: The Means to Success in World Politics*. New York: Public Affairs Press.

———. 2004. The decline of America's soft power. *Foreign Affairs* 83, no. 3: 16-20.

———. 1990. *Bound to lead: The changing nature of American power*. New York: Basic Books.

Ogoura, Kazuo. 2006. The limits of soft power. *Japan Echo* 33, no. 5: 60-65.

Price, Marie. 1994. Ecopolitics and environmental organizations in Latin America. *Geographical Review* 91, no. 1/2: 143-50.

Saxon, Earl, B. Baker, F. Hoffman, W Hargrove, and C. Zganjar. 2005. Mapping environments at risk under different global climate change scenarios. *Ecology Letters* 8: 53-60.

The Nature Conservancy. 2007. The Nature Conservancy annual report–2005. Washington, DC. http://www.nature.org/aboutus/annualreport/files/annualreport2007.pdf.

The Nature Conservancy. 2008. Climate change initiatives. Washington, DC. http://www.nature.org/initiatives/climatechange/. TNC staff. 2006-08. Series of telephone interviews conducted by Katrina Rogers.

United Nations Framework on Convention on Climate Change. 2008. Kyoto Protocols. UNFCC. http://unfccc.int/kyoto_protocol/items/2830.php.

US Congressional Legislation. 2008. Climate legislation side by side. US Senate. http://energy.senate.gov/public/_files/ClimateLegislationSidebySide110thCongress.pdf.

Young, Oran, and Marc Levy. 1999. The effectiveness of international environmental regimes. In *The effectiveness of international environmental regimes: Causal connections and behavioral mechanisms*, ed. Oran Young, 1-32. Cambridge, MA: MIT Press.

12

Climate Changes Everything

Dune Lankard

The events of March 24, 1989, when the Exxon Valdez tanker slammed into Bligh Reef, spewing more than 30 million gallons of crude oil into Prince William Sound, changed my life. I decided to become a full-time community activist and an uncompromising "voice of reason." On behalf of our Eyak people and also Eyak Corporation shareholders, I successfully worked through the legal process to preserve our ancestral rain forest and our subsistence and fishing way of life. Since then, I have continued working for social change, including having founded four not-for-profit organizations. And now we are organizing several for-profit sustainable and sensible businesses. The overarching goal is to permanently protect Alaska's wild salmon habitat in the Prince William Sound and Copper River watersheds, to provide education and awareness, and to explore sustainable alternatives. We as individuals and non-governmental organizations (NGOs) have to take action and work together to combat global climate change and save our planet.

Climate: *1: A region of the earth having specified climatic condition. 2 a: The average course or condition of the weather at a place usually over a period of years as exhibited by temperature, wind velocity, and precipitation b: the prevailing set of conditions (as of temperature and humidity) indoors <a climate-controlled office> 3: The prevailing influence or environmental conditions characterizing a group or period: <a climate of fear>* (Merriam-Webster 2009).

Introduction

My life, my world, was inexorably changed in one day. I put before you that these events and the one significant tragedy that caused this momentous change are analogous to what we are all experiencing now on this planet: climate change.

The History

After crossing glacial ice fields in the interior of Alaska, our tiny Eyak Athabaskan Nation would thrive over the next 3,500 years in small villages and subsistence campsites dotted along 300 miles of the Gulf of Alaska coastline. We certainly would have perished in the unforgiving coastal environment nestled along the Copper River Delta and eastern Prince William Sound had we not learned to adapt to harsh and changing climates.

The Eyaks lived along a thin green strip of forest and wetland habitat created from the receding glaciers. Our villages had to remain small in number because of limited habitable land and scarce food supplies. The force of Mother Nature also challenged the Eyaks with gale force winds and numerous torrential downpours of rain annually. We survived by knowing the seasons for hunting and gathering mammals, birds, fish, plants, and berries. Our people miraculously endured the tribal assimilation by the powerful warring Tlingit, Aleut, and Chugach Nations, and we survived the ruthless occupation of our homelands by the Spaniards and Russians. By the late 1800s, we were faced with social illnesses and issues—tuberculosis, small pox, and influenza, displacement of traditional lands, and non-Native oppression and suppression. We survived serious life-threatening attacks on our natural resources from seafood canneries, clearing of forests, road building, railroads, mining, and the near-annihilation of our natural food resources (razor clams, wild salmon runs, herring, and crustaceans). The Eyaks may have been erased from living memory had we not been the barterers and traders for tribes living along the coast and up the 300-mile Copper River.

The next round of attacks on the Eyaks, and indeed on all Native people of Alaska, was the erosion of our inherent sovereignty, subsistence, and spirituality through Alaskan statehood in 1959 (the year I was born) and the Alaska Native Claims Settlement Act in 1971 (ANCSA).[1] ANCSA abolished all aboriginal land claims; tribes and villages were forced to accept the mantle of the United States and create Native Corporations, or completely forfeit their land claims. In "lower 48" America, land "reserved" for indigenous peoples was set aside in reservations. In Alaska, however, ANCSA required that our lands be geographically defined and negotiated and turned into these so-called Native Corporations. Unbelievably, anyone who could prove they were at least a "quarter-blood" Alaskan Native could be a legal "shareholder" of the Native Corporation closest to where they lived in 1971 (ANCSA 1971).

Additional acts of Congress in the 1970s and 1980s further attacked our way of life. The Limited Entry Act of 1975 turned our fin fish (wild salmon) into a "commodity" by allowing our fishing rights to be bought and sold to anyone interested in buying a fishing permit, thereby allowing the ownership over fish resources. With that went our fishing freedom—our wild salmon subsistence. The Alaska National Interest Lands Conservation Act (ANILCA 1980) locked up over 100,000,000 acres of our ancestral lands in state and federal parks, which are now governed by the State and Federal government. This was followed by the 1986 Tax Reform Act (Tax Reform Act), which allowed the Native Corporations to create bogus "paper losses" called Net Operating Losses (NOL) and then sell those losses to U.S. corporations as easily abused tax loopholes. As a result, nearly a billion dollars of bogus NOL transactions instigated uncontrollable resource extraction projects that further separated the indigenous people from their traditional lands.

The history of our Eyak people may seem tragic, yet despite our small population and the challenges that we have faced, we remain alive, proud, and poised to offer what humanity we can to the world. One of the gifts that we offer is our spiritual belief that in order for the human race to survive, we need to transcend the legal, political, and economic sectors that keep us separated, confused, unorganized, and paralyzed in our efforts to protect our inherent rights as indigenous inhabitants of planet Earth, including the rights to clean air, clean water, productive lands, and thriving oceans.

The Problem

I grew up on the eastern shores of Prince William Sound (PWS), in Cordova, Alaska, in one of the last fishing families to make a living working together chasing wild salmon in the Copper River Delta and PWS. In my family, we were taught to be thankful and respectful of our environment and the precious remaining and recovering natural resources. Regardless, living from the land and ocean, we took for granted the incredible beauty and bounty of PWS and the Copper River Delta. We believed this amazing region would forever remain pristine and provide for our incomparable wild salmon way of life. We thought the wilderness and clean waters would always be a part of our way of life, as they defined who we were.

Our climate changed on Good Friday, March 24, 1989, on the twenty-fifth anniversary of the devastating Alaska Earthquake (centered in PWS), when the Exxon Valdez tanker slammed into Bligh Reef, spewing more than over 30 million gallons of crude oil into Prince William Sound. To me, that was "the day the water died." Strangely, something inside of me also came to life. I realized PWS would never be the same—that the entire world would now find out about our hidden paradise and that our worst fears about an oil spill occurring in our lifetime had in fact happened. Our reality and our dreams were forever altered on that day.

We looked around, hoping that the Coast Guard, the Alaskan government, the U.S. Congress, even our Alaska Congressional Delegation, or the Federal courts would force Exxon to make it right with the people, wildlife, and sea life in the "Exxon spill zone." Over 3,500 miles of coastline were drenched in crude oil as a result of this human-made catastrophe. Help, however, never came. The oil-spill cleanup was a farce—a media stunt to make the rest of the world believe that an oil spill of this magnitude could be contained and cleaned up. In truth, we were left to ourselves to defend and restore our environment and fishing way of life.

My sister Pamela was pregnant with her daughter, Jen, and working at our local Eyak (village) Corporation headquarters when the CEO emerged from his office and said, "*The whole world is busy watching the nation's worst oil spill. ... We need to start clear-cutting our forests as fast as we can!*" Our Native Corporation had decided to clear-cut our forests a couple years earlier to create a new cash flow for the corporation, and they were just starting to endure some heat from local environmentalists and fishermen. I was in Arizona at the time of the oil spill, pioneering a commercial fishery to remove "rough fish" (carp and buffalo) from lakes and sell them, rather than allow the government to continue using poisonous rotenone to eradicate these unwanted fish, when Pamela called. She said, "Dune, you need to come home now and fight for our forests or don't bother coming home, because there won't be anything left to come home to." I cried deeply, knowing that our wild salmon way of life and Eyak dreams and culture were on the verge of being wiped out once again. I knew I had to step forward and do something, anything, or forever wish I had.

I said my prayers and decided that I would have to become a "formless warrior" and "be louder than everything else" if I was going to have any chance of saving what was left of my homeland. Soon after becoming a fulltime community activist and an uncompromising "voice of reason," I was given my Eyak name by our honorary chief, the last full-blooded Eyak and only Eyak speaker of our language, the late Marie Smith-Jones—*Jamachakih*, which means "Little Bird (in the forest) who screams really loud and won't shut up."

I *knew* that if we allowed the clear-cutting of our old growth forests along the parallel path of the Exxon Valdez oil spill, the nation's worst oil spill, we would destroy our fisheries in this lifetime and possibly forever. Case studies clearly point out that wild salmon are "forest animals." They are born in forested lakes, travel down the rivers, and then head out to sea for two to seven years, depending on their species. If their once lush habitat has been destroyed when they return to "spawn and die," the salmon will spawn closer to cooler, deeper waters where the heavy rains, winds, and high tides wash the eggs away. Mortality rates will rise and the fish will disappear.

We saw that we needed to unite all of the stakeholders and stop the hemorrhaging of our wild salmon way of life by preserving our remaining threatened habitat and restore what was being destroyed by the spill. A handful of local, dedicated community activists came together, and we lobbied Exxon to settle its criminal case in an "out-of-court" settlement that established a $900 million dollar "restoration fund" to mitigate the impacts of the spill and help us preserve and restore what we could. We lobbied for those restoration funds to be used to buy the timber harvesting rights and development rights from the ANCSA corporations.

After a couple of unwarranted lawsuits against me, waged by the ANCSA-created Eyak Corporation, I filed a lawsuit on behalf of our Eyak people, and then again on behalf of our Eyak Corporation shareholders, aimed at protecting our inherent rights as tribal members and ANCSA shareholders. After not prevailing in the State Superior Courts, we appealed and won in the Alaska Supreme Court, which agreed that our lawsuits were in fact in "the best interest of the public." The Eyak Corporation then decided to not do battle in the courts and finally gave its 326 shareholders the right to choose and vote between preservation over development of our ancestral rainforest. Over 85 percent of the Eyak Corporation shareholders voted in favor of preserving the wild salmon habitat, followed by 85 percent of our ANCSA shareholders in our four Chugach region village corporations voting in favor of habitat protection.

After that pivotal vote, a dozen more Native Corporations in the Exxon spill zone engaged in land conservation negotiations, sparing their forests from the saw. Again, more than 85 percent of these Native shareholders voted in favor of conservation. A conservation precedent was set, with Native shareholders receiving over $400 million from the out-of-court restoration settlement to leave upwards of 700,000 acres of their wild salmon habitat intact. Indigenous people were finally allowed to determine their own fate, thereby preserving their subsistence and commercial fishing way of life, hopefully for generations and generations yet to come.

Although this bold and risky strategy was successful and precedent-setting for conservation, it only helped stave off unsustainable development of threatened salmon habitat in the spill zone. Today, Native corporations, industry, and government continue to push for unsustainable development, via roads, trails, oil and gas drilling, coal mining, and industrial tourism, all of which will only lead to allocation wars, overfishing, and overhunting, followed by uncontrollable access to the Copper River Delta, PWS regions, and throughout Alaska itself.

An example of an unsustainable development proposal is the Bering River Coalfield, which lies fifty-five miles east of Cordova on the eastern edge of the Copper River Delta. Numerous Bering coal appraisals show that there could be a minimum of 35 million tons of harvestable coal and upwards of three billion tons in the Carbon Mountain region. There has been a push by our regional Native corporation, the Chugach Alaska Corporation, and the Korean owners of the Bering River Coalfield to open a road crossing dozens of wild salmon streams across the Copper River Delta. This area is rich with fossil fuel resources—oil, gas, and coal—and boasts several hundred thousand acres of old-growth trees. For the last 100-plus years there has been a visceral drive by political and industry-minded carpetbaggers to open this pristine roadless wilderness area to massive unsustainable development.

Why would one ever think of sacrificing what is sacred, thriving, renewable, and sustainable for unsustainable short-term greed? If these resources are ever extracted, our magnificent glaciers will melt even faster than they already are. And glacial runoff water is what makes up the 300-mile Copper River drainage system. This is where our wild Copper River salmon return each year to spawn and regenerate. Case studies prove that when wild salmon spawning habitat is developed, even by 10 percent, wild salmon runs retreat and disappear.

The Solutions

It is quite clear to me that how we choose to manage our resources in the Copper River Delta and Prince William Sound is a microcosm of what is wrong and what is right in the world.

For us, the wild salmon are the symbol of everything we are and everything we have. They *are* at the center of our spiritual, subsistence, and sovereign way of life. Our overarching goal here is the permanent protection of the habitat—to build unity and support for "Wild Salmon

Preserve" legislation so the stakeholders who live here and fish here and who are dependent on our wild salmon take part in their fate. Only about 10,000 people live in the thirty-some Copper River and Sound communities. We plan to implement a locally based, yet international in scope, "Copper River Wild Salmon Forever" campaign to jumpstart this legislative push to protect our wild salmon and spawning habitat. Who can argue with Wild Salmon Forever?

We need to step up global efforts to come up with local, national, and worldwide solutions addressing climate change, because climate change changes everything. We must build alliances and unity, starting from a grassroots level that calls for permanent protection of our planet's remaining wild places, while also demanding full accountability and responsibility from multinational corporations and world governments.

Author Paul Hawkins wrote in *Blessed Unrest* that there is a worldwide "Climate Change" movement out there, but that it has not yet been identified by any unifying approach. A dear friend and professor extraordinaire, Mary Wood from the University of Oregon, has written a paper called "Nature's Trust: Reclaiming an Environmental Discourse" (Wood 2007, 243). Nature's Trust is a unifying approach, as it has the potential to integrate law (trust obligation), spiritualism (duty to future generations), and economics (natural capitalism). Professor Wood says that these principles reflect, in their broadest form, the approach of Native societies, which are the only human societies that have successfully integrated all of the aspects of human behavior (economy, governance, law, religion, and spiritualism) (ibid.).

Professor Wood has proposed a legal climate change strategy, called "Atmospheric Trust Litigation," in which she characterizes the Earth's atmosphere as an asset that the people of the world own in common. This approach has a basis in our environmental law. She says, "You can even think of it as an attribute of sovereignty, the duty of government (as a trustee of that asset) is to protect our natural resources. The atmosphere is the most crucial resource in our trust, because it holds everything else together." Professor Wood further says, "Global warming is an urgent matter. This approach underscores, or brings to light, an organic duty on the part of virtually every level of government [in the world]" (Wood June 24, 2009).

> Friend and associate Professor Richard "Rick" Steiner has come up with a visionary idea to implement a one percent Earth Profits Fund, a private-sector global conservation finance initiative for the twenty-first century that would provide increased and sustained support to biodiversity conservation and sustainable development. "As

envisioned," Professor Steiner says, "participating companies would dedicate 1% of their after-tax profits annually into a pooled Fund, which would support urgent priorities of habitat conservation, Red-List species protection and recovery programs, and other biodiversity conservation and sustainable development projects throughout the world" (Steiner 2005, 1).

Professor Steiner states that past combined financial commitments of governments, international financial institutions, corporations, and private philanthropies have not been sufficient to reverse environmental decline (ibid., 2). Some estimates suggest that as much as $300 billion (USD) per year would be necessary to fully protect the global environment, while others suggest that at least $30 to $40 billion per year in dedicated funding would be necessary (ibid., 1). Set against these estimates, global annual spending specifically for conservation is perhaps just $3 billion to $7 billion per year (ibid.). Whatever the estimates may be for current conservation expenditures vs. need, all agree that there exists a critical shortfall and urgent need to substantially increase conservation finance (ibid.).

Data published by Dr. James Hansen of the NASA Goddard Institute for Space Studies show that CO2 emissions are not the main cause of observed atmospheric warming (Hansen 2004). He stresses that the most important non-CO_2 greenhouse gas is methane, which comes primarily from animal agriculture (ibid., cited in Mohr 2005, 3). Dr. Hansen further says that since methane emissions cause nearly half of the planet's human-induced warming, it must be a priority to reduce methane emissions (Hansen 2004, cited in Mohr 2005, 3). In a recent paper entitled "A New Global Warming Strategy" (Mohr 2005), Dr. Hansen is cited extensively for the argument that the number one source of methane emissions worldwide is animal agriculture, and that this industry produces more than 100 million tons of methane a year. The conclusion is that one way to greatly reduce global warming in our lifetime is to reduce or eliminate animal products from our diets. The paper further states, "[U]nlike carbon dioxide, which can remain in the air for more than a century, methane cycles out of the atmosphere in just eight years, so that lower methane emissions quickly translate to cooling of the earth" (Mohr 2005, 4). Changing our diets may seem drastic or unreasonable to many, even to inhabitants of Alaska, but it will take drastic measures to limit global warming as millions of acres of land around the world are currently being cleared for animal agriculture. This negative development impacts traditional indigenous peoples around the world, as indigenous people are displaced and their traditional lands destroyed and wild foods replaced with processed foods that lead to diabetes. Let us be clear: Diabetes does not just impact indigenous peoples, it impacts everyone. One person dies from diabetes every ten seconds. This is inhumane and preventable.

Eating less meat goes hand-in-hand with promoting healthy diets overall. To this end, organic farming is a necessity: Organic farms are healthier for the land, for crops, and for people. And if indigenous peoples are encouraged to grow their organic traditional foods, and once again live from the land and the water, they would be healthier and more vibrant—and less likely to drink alcohol or to use drugs. There would be more pride, self-esteem, and cultural revitalization worldwide.

These are the kinds of bold and visionary climate change strategies that we, as a human race, need to consider, promote, and support immediately, if we are going to have any chance of success in preserving or protecting our human existence.

What is next is water. Allocation wars over water are not new, and yet having access to fresh, clean drinking water is increasingly becoming a critical worldwide issue. First oil, now water—we are depleting our world resources at unprecedented rates. If we start *today* to conserve electricity and properly manage the natural resources we have left, we will not have to pay more for water than we do for fuel. We are smarter than this, are we not?

Much like the slogans, "Get Out the Vote" and "Think Globally, Act Locally," and even the public media campaigns in which tobacco companies educate the public as to the real harm of cigarettes, doing so in the wake of their lies and unlawful actions, we need to start a "Climate Change Campaign" as a "call to arms" to address, educate, and provide solutions for humans on global warming issues. It all starts with us.

We are doing our part here in Alaska. We have created a simple new equation that we live by: *non-profit + for-profit = social profit*. We realized that in order to continue our work as "social change artists," especially in times of hardship, we need to be more self-reliant and not totally grant dependent.

We have founded four non-profits to create change:

- EPC (Eyak Preservation Council), a cultural and environmental preservation non-profit addressing indigenous, habitat, and restoration issues in the Copper River and Prince William Sound region (www.redzone.org).
- NATIVE Conservancy Land Trust, an indigenous-led organization that helps us protect not only endangered plants and animals, but endangered peoples and their culture (www.redzoneorg).

- FIRE Fund (Fund for Indigenous Rights and the Environment), an endowment allowing us to direct energy (time, money, or love) to what does the most good in the shortest amount of time (www.firefnd.org).
- RED OIL (Resisting Environmental Destruction On Indigenous Lands), a statewide chapter organization whereby we direct energy to stop all new oil and gas projects and bad development projects that are detrimental to our unique subsistence and traditional ways of life (www.ienearth.org/redoil.html).

We are in the formation stage of one more social-profit: the Cordova Community Cold Storage & Cookery (CCCSC). CCCSC will help our community improve the quality and quantities of our local subsistence and commercial fishing foods. The proposed building will offer three separate services: one where commercial fishermen can process, value-add, and direct market their catch; another for subsistence users, where they can skin, process, steak, package, and blast-freeze their fish and game animals; and an area for research and product development, thus jumpstarting a local cottage industry. The CCCSC plant will be designed to take advantage of renewable sources and upcoming energy technologies specific to rural Alaska—summer sun-solar; winter-wind; tidal, hydro, geo-thermal, and biodiesel developed from fish offal (waste).

Additionally, we are now organizing several for-profit sustainable and sensible businesses:

- Copper River Wild Salmon Company, which will help fishermen improve the quality of their catch by promoting at-sea processing, value-adding, and direct marketing of their seafood catches—therefore bringing higher-quality products to consumers at lower prices than retail, while bringing higher prices paid to fishermen.
- We are negotiating with Steyr Motors (Austria) for an Alaska dealership. Their state-of-the-art marine diesel engines are biodiesel compatible and can drastically improve fuel economy for fishing vessels. We installed two Steyr engines in one of our partner boats, the engines burned fifty gallons for each 24-hour fishing opener, compared to 120-200 gallons consumed each opener by every other vessel. Each engine will save over $10,000 of fuel bills annually.
- We are also in the planning and building stages for three eco-lodges on the Copper River Delta and in Prince William Sound. We have many visitors each year, and we want to provide them with first-hand experiences in the Delta and PWS, as well as to educate and enlighten them in regard to our Wild Salmon Preserve legislation campaign.
- Education and experience are crucial. Therefore, we are also planning to build a progressive hands-on school at Knight Island in PWS that

> will focus on eight key areas to help people empower themselves in their respective homelands: oil spill response and restoration; fishery management and restoration; implementation of renewable energy projects; environmental law; wilderness survival (including learning how to subsist in the wild); understanding the public process—how to run for city council, write grants, lobby the legislature, or run for President of the U.S.; teaching and providing information on the Cultural Conservation Initiative, for which I won an ASHOKA fellowship (teaching people how conservation is more valuable economically than short-term development of our natural resources); and designing a world-class retreat center for rest, recharging, and revitalization in these chaotic and changing times (Ashoka).

These organizations and campaigns will work hand-in-hand to complement our over-arching goal for permanent protection of the habitat and to provide education, awareness, doable alternatives, and modern-day models that can be duplicated around the world.

The Deal

For the record, I do not have a lot of faith in pumping chemicals or carbon into the ground as a way of truly limiting or eliminating harmful gases and pollutants from getting into our air. Yet I do believe that there is merit in creating a "carbon credit" program through which private landowners and businesses are encouraged financially to sequester fossil fuels by leaving them in the ground, never to be drilled, mined, or emitted into our atmosphere. This is what we want to do with the Bering River Coalfield in the eastern Copper River Delta, once we are successful in our efforts to purchase the coal patents.

A carbon credit program could work if the landowners who retire their fossil fuels receive payments contingent on furthering critical renewable energy research and development and/or the financing of projects that help our planet heal and that help communities become fossil fuel-free and self-sufficient—projects like natural energy (solar, wind, tidal, hydro, hydrogen, organic bio-fuels, etc.) and conservation efforts to preserve or restore vast regions of trees and habitat that help filter air and water. We must change our view and standards of status quo economics and be more creative and proactive in designing and implementing programs that help communities foster and develop sound, viable, sustainable community and regional approaches to services and local economies.

To move things forward, indigenous tribes around the world, including U.S. tribes, should consider signing the Kyoto Protocol as sovereign Nations, so they can make unilateral decisions about global climate change.

The Third World debt situation, along with crashing economies and environments, is a good example of what has gone wrong in the world: poor countries forced to destroy the only habitat they have—endangered habitats that the world needs intact—to pay off their global debt. We need to work together as one living planet.

Indigenous peoples have clocked thousands of generations of good stewardship on their lands, with only a few decades of being misled by globalization and promises of a better lifestyle. As Professor Mary Wood says, "We need to find a way to once again integrate spiritualism and capitalism into law" (Wood June 24, 2009). Remaining in denial, being complacent, or staying frozen in a state of panic will do nothing but exacerbate our already dire situation.

Another good example is the current U.S. and worldwide financial crisis. We need to stop the pillaging and foreclosures by Wall Street and the banks; we must demand and see that our new government implements long-term strategies that help everyone, not just "big money" and the CEOs. We must support President Obama and the new administration during these challenging times, while calling for an emphasis on public oversight and citizen-friendly policies that protect our personal investments, with strict regulations and penalties placed on corporations and banks to ensure that a crisis of this magnitude does not occur again.

Let us start by taxing what is bad and promoting what is good for the planet. Implementing a "global fossil fuel tax," or endorsing the one percent Earth Profits Fund, for example, would create billions of dollars in annual revenues for conservation efforts aimed at protecting and promoting sustainable and renewable energy economies and for the building of sustainable, resilient communities. ASHOKA Fellow Bruce Cahan has come up with a wonderful plan to create "a sustainable resilient bank" called the "GoodBank" that would fund communities with lower long-term interest rates if they can prove that their community projects are well thought-out, intentionally built, and can sustain themselves during hardships such as earthquakes, tidal waves, oil spills, etc. (Ashoka). Communities would then have to think harder and be wiser about how best to take advantage of renewable energy resources and how to develop sustainable communities and economies. The payoff would be great and last forever.

Another example of sustainability in coastal communities, where fishing is the dominant and most sustainable economy, is to privatize fisheries management and provide set allocations to help shore up endangered stocks. Fishermen who have more of a stake in management and ownership will take more pride and effort in protecting resources, ocean habitat, and limiting harvests. Many of the planet's fisheries are in a state of peril because of mismanagement, no management, overfishing, and pollution, and they are in dire need of restoration and preservation. Restoring the fisheries does not mean that these problems will not occur again, but a long-term, hands-on private management approach by active stakeholders will lead to recovery and the ability to make a sustainable living and feed people worldwide from our oceans once again.

Here in Alaska, global climate change means our glaciers are melting faster than ever before, beetles are rampant and destroying our old-growth forests, and the permafrost is melting at unprecedented rates. Storms, flooding, and rising oceans are wreaking havoc on our coastal villages and wild salmon runs, and polar bears and walruses have less and less ice to play and hunt on. Where does it end? And where does stopping it begin? What is threatening our communities and our sustainable economies, including our precious environment, is our refusal to take positive action *now* to change the path we are on. It will take all of us, as current stewards of the planet that we all call home, working together to regain control of our destiny, especially if we want to give our children a sporting chance of survival.

The role of ethical transformation, among the general population and within NGOs and non-profit organizations, happens when people decide enough is enough. We have reached the point of no return. It is true that the government, civil society, and the economy are all part of a system that is interconnected and interdependent, but it is we humans who ultimately have control of our fate and the responsibility to rewrite this next chapter for our planet.

We possess the knowledge and experience to know what to do, but can we spark the courage and the political and moral will to do the right thing? In truth, we do not have a choice. As Henry David Thoreau wrote, "It takes two to speak the truth: one to speak, and another to hear." This statement speaks volumes to what we need to do as a human race. We all know there is an illness spreading around our planet; we need to listen to our Mother Earth's cry for love, to be present, and find the courage to take action while we can still make a difference.

A Brief Summary about the State of Alaska and the Alaska Native Claims Settlement Act

Much has happened to Native Alaskans in the last 250 years since European contact and occupation. Over the years, the cultures, languages, and subsistence lifestyles of some Alaska tribes have remained somewhat intact, although threatened. Others, like the Eyak, who lived along the Copper River Delta and eastern Prince William Sound rainforest, were devastated. Traditional Eyak village sites were torn down to build the new foundation for the City of Cordova. Cordova grew rapidly in the late 1880s because of fish traps and salmon canneries. There was another growth surge in the early 1900s following the discovery of a huge pure copper vein in the Copper River/Kennicott region. When Alaska became the forty-ninth State in 1959 there were approximately 85,000 Indigenous people surviving throughout Alaska.

In the 1960s, oil was discovered on the North Slope of Alaska. Congress immediately was faced with another "Indian problem," namely, how to put an 800-mile pipeline through all those Native villages when the rights of title in Alaska were already truly obscure. A lawsuit filed by an Eskimo man by the name of Charlie Edwardson held off oil development from 1966 until 1971. In 1971, the Federal Government, with pressure and $562 million dollars from the oil companies, decided to take things into their own hands. President Nixon convinced Walter Hickel, Governor of Alaska, to retire and immediately be appointed as the new Secretary of the Interior in order to broker the Alaska Native Claims Settlement Act (ANCSA 1971).

All aboriginal land claims were extinguished by this act of Congress, which was passed without a vote from Alaska's original inhabitants or the American public. Alaska Natives retained only 44 million acres of land (approximately 11 percent of 380 million acres) and $962 million dollars. These lands became for-profit Native corporations. Natives in Alaska do not own their ancestral land, they own shares in an ANCSA corporation that holds land title.

One hundred shares were given to every Alaska Native alive in 1971 who could prove they had a quarter-blood of Alaska Native heritage and lived in a local traditional village. No additional ANCSA shares were to be issued to their children. These ANCSA shares cannot be sold or traded, but only transferred to a relative by inheritance. ANCSA corporations are not mandated to have anything in their charters regarding culture, subsistence, or ancestral land preservation, but simply to be for-profit corporations. If they are not profitable, the Federal Government can take possession of the ANCSA land and dismantle the corporation. Some have gone bankrupt or merged with other ANCSA corporations. The majority of ANCSA corporations attempt to make profits by extracting and exploiting their natural resources.

It is only a matter of time that due to dilution of ANCSA shares, both in the number of shares owned and because of lack of Native blood-quantum requirements, there will be no more Native Corporations, except possibly in name only. We believe that preserving ancestral lands will forever preserve the Native spirit, culture, and subsistence way of life for those who chose to live from the land and the ocean. ANCSA corporations, and all tribal entities, are faced with how they will create financial revenues in order to provide for services and dividends for their Native people, who become more dependent on the culture of money and Western materialism every day.

References

ANCSA 1971, 43 U.S.C. § 1601, *et seq.*

ANILCA 1980, 16 U.S.C. § 3101, *et seq.*

Ashoka.org/fellows, accessed August 2009.

Hansen, James, Inventory of U.S. Greenhouse Gas Emissions: 1990-2002, U.S. EPA, April 15, 2004.

Merriam-Webster Dictionary. Available at http://www.merriam-webster.com/dictionary/climate, accessed August 2009. Mohr, Noam. A New Global Warming Strategy: How Environmentalists are Overlooking Vegetarianism as the Most Effective Tool Against climate Change in Our Lifetimes, EarthSave International Report, August 2005. Available at http://www.earthsave.org/news/earthsave_global_warming_report.pdf.

Steiner, R.G. The 1% Earth Profits Fund: A Private Sector Conservation Finance Initiative for the 21st Century. IUCN Commission on Environmental Economic and Social Policy (CEESP) / Social and Environmental Accountability of the Private Sector (SEAPRISE) website, 2005. Available at www.iucn.nl/sbeos/doc/file.php?nid=6574.

Tax Reform Act, 26 U.S.C. §§ 1042, *et seq.*

Wood, Mary. Nature's Trust: Reclaiming an Environmental Discourse," 25 Va. Evtl. L.J. (May 2007). For further reading on this approach, see "Advancing the Sovereign Trust of Government to Safeguard the Environment for Present and Future Generations (Part I): Ecological Realism and the Need for a Paradigm Shift," 39 Envtl. L. (2009) p. 43.

Wood, Mary, email to the author, June 24, 2009.

Part V

Case Studies

13

Trends and Impacts of Climate Change in Cameroon, Central Africa: Considerations for Renewed Ethics towards Resilience Options for the Community

Samuel Ayonghe

Cameroon, a country situated in Central Africa, is endowed with varied ecological diversity that ranges from coastal mangroves and rich Equatorial rainforests in the south, through Savannah and Sahel ecosystems, to the southern fringes of the Sahara desert around Lake Chad in the north. It could accordingly be considered more or less as representative of Africa in miniature. An evaluation of parameters associated with climate change and its impacts in the country based on existing data on the subject during the last century (1926 to 2007) is indicative of increasing trends with national magnitudes similar to those of global trends. Regional trends of climate change across the country are assessed, their impacts on vulnerable communities evaluated, and possible mitigation and adaptation approaches discussed with specific reference to ongoing research and outreach activities. Recent proactive government policy and action to create a national observatory for climate change in the country is presented. The status, mission, composition, and anticipated activities of such an observatory are discussed with the aim of ensuring its scientific rationale, as well as the social, economic, ethical, and cultural implications of decisions to be considered by the government in its policy towards redressing issues on climate change through such an observatory. Discussions of some ongoing policies, capacity building, and sensitization experiences at the national and international levels, especially on the African continent, are used to highlight the advantages and pitfalls to be considered when providing the most optimal approaches of tackling this human suicidal phenomenon, which is already decimating humankind and other species on the Earth's surface at an alarming rate. Renewed environmental and political ethics on the subject are proposed for future consideration.

Introduction

It is a known fact that climate change is having a drastic impact on the environment and the population globally. According to Nobel Laureates (2001 in McBean 2004, 183), 110 laureates in 2001 identified both climate change and a weaponized world as the two most important challenges facing humanity in terms of international security. In fact, climate change has recently become a serious threat to human survival on this planet and is accordingly an important issue of debate by politicians on economic, developmental, and socio-cultural issues. Controversies about its natural and/or anthropogenic causes, its impacts on the society, and future projections abound. The truth, however, is that the rate of global warming and its impacts on climate change across the entire globe during the past two to three decades is unprecedented in recent and even geologic history as seen from extreme warming rates, sea level rises, droughts, heat waves, depletion of water resources, sporadic precipitation events with accompanying floods and landslides, and hurricanes, among others.

There are also constant changes in the social and demographic patterns of communities in the developing world, with high population growth in vulnerable areas, as well as with a growing inequality between the richer and poorer sections of society, with the latter being more vulnerable to climate change-related disasters.

There is, however, no effective and continuous monitoring and evaluation of the trends and impacts of the phenomenon in most developing countries aimed at sensitizing the community on mitigation and adaptation options. This is due to inadequate infrastructure and a lack of innovative tools and methods, relevant expertise, incentives, and more importantly, government policy favoring holistic approaches for evaluating the phenomenon. Improved systems for monitoring and reporting extreme weather events linked to climate change are available in the developed world, while developing countries such as Cameroon are affected without receiving any warning signals.

International conventions relating to the impacts of climate change since the early 1990s have been limited to water issues, land degradation, and biodiversity loss, with guidelines developed to be undertaken by the national partner countries based on global trends without due consideration to the exact regional/local trends and their associated impacts, especially in developing countries.

Greenhouse gases (GHGs) have been identified by scientists as the main cause of global warming, which is intrinsically linked to climate change, but regrettably, the 1992 UN Framework Convention on Climate Change, the subsequent Kyoto Protocol, and Bali Roadmap, all aimed at reducing the concentrations of these gases to acceptable levels in an attempt to reduce the impacts of climate change, are hardly respected by the highly industrialized nations responsible for producing the highest amounts of these gases. As observed by Lagos (2008, 19), due to the limited calendar and low participation, as well as the insufficient follow-up in its implementation, the Kyoto Protocol has never been considered as a solution to climate change. Some countries failed to ratify or accede to this Protocol, while those that did so have not abided by its objectives.

Since the developed countries have been responsible for the accumulated concentrations of GHGs over the decades, they are accordingly expected to be the first to react by providing financial and technological aid to the underdeveloped countries in order to assist them in the implementation of strategies. They are thus expected to spearhead all mitigation actions, but their reluctance to abide by established ethical norms could lead to disastrous impacts such as the melting of polar ice sheets in Greenland and in the Arctic and Antarctic regions, which will lead to the flooding of coastal cities where the highest concentrations of the world's population are located, among other impacts.

Although the developing countries, such as Cameroon, are the least producers of GHGs (mainly from activities associated with deforestation), they are the most vulnerable to this phenomenon even if provided with finances accrued from the universal carbon taxes proposed by Sir Nicolas Stern in 2006. This is due to the inexistence of appropriate structures to handle such funds as well as the exponential growth of vulnerable populations that depend on subsistence agriculture, through deforestation in some areas. The change from such agricultural practices will involve the disruption, or a complete destruction, of socio-cultural practices and values, with accompanying new crops. The design of future adaptation options based on new technologies and approaches of research and developmental options will be necessary if the developed countries continue to emit high GHGs, since they will be expected to pay carbon taxes up front for their undesired actions.

There is still a huge variation in the scientific estimates of the percentage by which the climate will warm in future. On a global scale, current climate prediction models utilize a wide range of data using GHG emissions, volcanic activity, ocean components, etc. Such predictions are carried out either using current data from the atmosphere or by working with data collected over the years to produce models which go up to 2050 and even beyond.

There is thus cause for concern about mitigation and/or adaptability approaches to be considered in order to redress this ominous global catastrophe. This chapter is therefore focused on assessing the exact pattern of trends and impacts of climate change in Cameroon from 1926 to 2007 and their relationship with sub-regional and global trends. The main objective is to utilize the results for the identification of appropriate mitigation and adaptation options to be adopted towards increasing the resilience of vulnerable communities based on the conception of sustainable development approaches in the most affected areas/regions of the country.

The absence of harmonized research and outreach activities on the subject, as well as ongoing policy options of the Cameroon Government towards setting up a national observatory for climate change across the country, which was based on the President's declaration at the Sixty-second Session of the United Nations in New York in September 2007, are discussed here. Special attention is paid to possibilities of defining or conceiving approaches of harmonizing multidisciplinary efforts in collaboration with international collaborators towards coping with the trends and impacts of the phenomenon. Outreach approaches through lectures and the use of communications guides are aimed at empowering Members of Parliament and Councillors on this phenomenon. The introduction of issues on climate change into compulsory university courses, capacity-building strategies on the continent by the United Nations Environmental Programme (UNEP), and the scientific research option offered by the International Council for Science Regional Office for Africa (icsu-africa.org)—all aimed at educating future generations on the state, trends, and impacts, as well as possible mitigation and adaptation strategies—are also discussed.

Trends of Climate Change in Cameroon in Comparison with Global Trends

Previous work on the variability and/or trends of climate change from 1930 to 1995 on water resources, in volcanic terrains, and on the environment, indicated linear increasing trends in temperatures and decreasing trends in total amounts of annual rainfall and number of days of rainfall (Ayonghe 1998, 159). The magnitude of the results correlates with global trends of a rate of temperature increase of 0.91°C during the twentieth century (Ayonghe 2001, 150).

Reliable global weather records in most developing countries have only been kept during the last hundred years or even less and it is only this recent climatic fluctuation which can be investigated adequately. In Cameroon, such records go back to 1926 (Ayonghe 2001, 143), and can be used to obtain trends, especially if taken over a period of more than thirty to thirty-five years.

Lean and Hinrichsen (1992) utilized records of weather conditions from 1861 to 1990 to show a global temperature rise of between 0.3°C and 0.6°C. Projections from their data into the future indicated that the world will be on average 1.3°C and 3°C warmer in 2020 and 2070 respectively, thus making the world hotter than it has been for the last two million years. Past changes of this magnitude of global warming took thousands of years and species could adapt slowly. If the present trends continue undisturbed, the effects will be disastrous even within the next twenty to thirty years; thus the need to act now.

I have written previously that there was a higher net rate of increase in temperature in Cameroon of 0.19°C from 1930 to 1995 as compared with the twentieth century warming rate of approximately 0.5°C from 1900 to 1991 (Ayonghe 2001, 150). A predicted temperature increase of 1.8°C by 2060 from the Cameroon data is far less than the global 3°C predicted by Lean and Hinrichsen (1992) for the year 2070. This is due to the use of a linear projection without due consideration of other parameters, whereas according to McBean (2004, 183), such projections should include assumptions regarding future emissions of greenhouse gases and aerosols, which are all dependent on industrialization and other technological changes, and their utilization of energy as well as the ever-increasing needs of an exponential population growth within the next few decades.

According to Pachauri (2008, 12), global temperatures increased by 0.74°C and the sea level rose by 17 centimeters during the twentieth century. This was accompanied by extreme precipitation, heat waves, floods, and drought. He further adds that the magnitudes of all these changes are still increasing and that we could lose 20 to 30 percent of the world's species if the temperature increases from 1.5 to 2.5°C and more. New varieties of crops that are more resistant to drought and extremely high temperatures could be tried as a form of adaptation under such conditions (Pachauri 2008, 13, 15).

In fact, climate change scientists at the Meteorological Office in the Hadley Centre and at the Climatic Research Unit at the University of East Anglia have maintained that global temperatures from 2000 to

2008 are almost 0.2°C warmer than the average for the decade 1990 to 1999, with the year 2007 being the second warmest year in history (Allen 2008, 1).

Sacchetti (2008, 20), however, presents contradictory views of skeptics on climate change who are of the opinion that, although 1990 to 1999 was the warmest decade ever recorded, with 1998 being the single hottest year yet measured, levels of CO_2 emissions have since then increased by four percent without an accompanying warming trend. He further states that measurements of global temperatures are faulty and do not indicate any increases and that, in a historic context, world temperatures have always varied and there should be no concern about present trends (Sacchetti 2008, 20). But he concludes that an infinite number of factors responsible for global warming can be invoked in addition to those cited by skeptics such as ocean currents, water vapor, and celestial phenomena (Sacchetti 2008, 21). The trends of disasters and catastrophes associated with climate change (hurricanes, cyclones, floods, droughts, heat waves, etc.) are, however, not reflected in their views.

Analysis of trends in rainfall across the country during the same period indicated a net reduction of the annual number of rainy days by seven days, while the total amount of annual rainfall dropped by 282 millimeters, equivalent to 43 millimeters per decade (Ayonghe 2001, 148). The decrease in rainfall over this period was more severe in the northern Sahel region of the country. Odjugo (2005, 145) described a similar decrease in both the total amount of annual rainfall and number of days of rainfall in Nigeria, based on data spanning from 1970 to 2002, and further stated that the phenomenon has created an ecological destabilization in the country by altering the pattern of vegetation belt, especially in the northern fringes of the country where wind erosion and desertification have been enhanced.

Impacts of Climate Change Globally and in Cameroon

Hazardous events related to climate change have been responsible for enormous economic costs around the world. These have increased from an average of two per year in the 1950s, and five per year in the 1970s, to nine per year in 1990s, with windstorms and floods accounting for about two-thirds of these catastrophes, and with about 10,000 and 25,000 lives lost in 2000 and 2001 respectively (McBean 2004, 178). The year 2003 was one of the warmest years on record with high temperatures recorded in Australia, South America, Canada, and parts of Africa. A record summer heat wave affected Europe, where the maximum temperature ever recorded in the United Kingdom was broken when it reached 38.1°C at Gravesend in Kent. In France, 14,802 people died from heat strokes, while in India, temperatures of between 45° to 50°C killed more than 1,500 people.

According to Pachauri (2008, 12), during the twentieth century, the average warming was 0.74°C and the extent of sea-level rise was 17 centimeters. The melting of the Greenland and the West Antarctic ice sheet will lead to a sea level rise of several meters and an increase in temperature of 1.5 to 2.5°C, and if the temperature rises further, it will lead to the extinction of 20 to 30 percent of the world's species, as mentioned above (Pachauri 2008, 15).

An evaluation of the principal impacts of climate change in Cameroon is indicative of an exponential increase in temperatures resulting in:

> The drying up of springs and streams and a reduction in the volumes of rivers, leading to the scarcity of potable water, with a resultant decrease of water to run the three hydroelectric power stations that supply electrical energy to the entire country and to the neighboring Republic of Central Africa (Ayonghe 2001, 152).
>
> The depletion of habitats (ecosystems) with the extinction of plant species and the biota that depend on them, resulting in droughts, desertification, and famine (Ayonghe 1998).
>
> Sporadic and prolonged precipitation leading to disastrous floods and landslides, resulting in deaths and the destruction of forests, property, and structures (Ayonghe et al. 1999, 13, Ayonghe et al. 2004, 438, Ayonghe & Ntasin 2008, 195).
>
> Migration of communities southwards from drought-stricken areas to greener areas (Ayonghe & Amawa 2007, 22).

Other observed impacts include: prevalence of infectious diseases such as malaria in new areas which were previously not infested by vectors of the disease; changes in agricultural patterns with reduction of crop yields due to increased temperatures and decreased rainfall; development of new plant strains with increased yields in some areas; reduced livestock in depleted habitats due to decreased rainfall, and deaths related to heat stress; changing seasonal patterns or shifting seasons accompanied by shifting vegetation zones in response to shifting isohyets; and increase in energy demands for cooling during high temperatures.

Mitigation, Adaptation, and Resilience Options

The establishment of a system of harmonized universal carbon and even general pollution taxes could reduce emissions and generate financial resources for developing clean energy sources, especially for developing countries where the poor are the most vulnerable and the least able to adapt. Since the increasing pattern of climate change is inevitable, the society which is most vulnerable to it will have to devise adaptation and resilience options in order to cope with this phenomenon. These should be based on education of stakeholders and the population and through the formulation of policy/regulation based on appropriate decision-making approaches by both the stakeholders and the vulnerable population. All these require investments which, in most cases, are not usually considered as a priority by the governments of developing countries, which usually only provide funds, backed by foreign aid, after the destruction produced by the climate change-related disasters. In fact, the provision of funds and materials towards research, monitoring, forecasting, and development of early-warning systems will pay dividends if well managed.

Towards Renewed Environmental and Political Ethics in Cameroon and on the Continent

Results of useful research efforts by experts on climate change and its impacts on the community in Cameroon are dispersed within different university institutions, ministries, and non-governmental organizations, as well as in the form of publications in various libraries abroad. The collation and harmonization of these efforts will be vital for conceiving and implementing further work on the subject, aimed at mitigating its consequences on the population, on the economy of the country, and on adaptation strategies. Outreach mitigation activities being implemented (such as the current tree-planting drives) are carried out uncoordinated and without due consideration on the future scenario of the trends of climate change.

The Ministry of Environment and Protection of Nature is currently organizing the creation of a National Observatory for Climate Change in compliance with the solemn announcement on this subject in September 2007 by the Head of State, His Excellency President Paul Biya at the 62nd General Assembly of the United Nations in New York. An Inter-Ministerial Committee was accordingly created on November 5, 2007, charged with the responsibility of defining the terms of reference of such an observatory, with the following issues to be considered in its anticipated mission:

1. The collection and analyses of data through continuous monitoring and the prediction of negative indicators of climate change through the utilization of

existing national and international organizations and appropriate computing technologies and data such as satellite images.
2. The formulation of a political orientation based on precise indicators towards mitigation against the impacts of climate change on the social and economic sectors of the country.
3. The conception of approaches of dissemination of results thereof, and the design of a plan of action towards sustainable development options.
4. The conception of ways of facilitating the restoration and conservation of the rich Equatorial forest ecosystem in the country and the reduction of GHGs in relation to the carbon tax.
5. The conception of approaches of collaboration with other observatories and the international community on the subject.

This observatory is intended to work in collaboration with representatives from the Prime Minister's Office, the Ministries of Environment (as the host), Research, Finance, Economy, Forestry, Agriculture, Animal Husbandry and Fisheries, Water & Energy, Transport, Public Health, and Higher Education (Universities), as well as with national and international non-governmental organizations, and other observatories. The range of its missions and partners will make it a multidisciplinary organ with diverse competences that will require varied material, human, technological, and financial resources to facilitate the realization of its missions under a central unit responsible for evaluating the scientific aspects for defining and streamlining the political orientation of the Observatory. It is expected to be charged with the responsibility of defining the politics and functioning of the observatory and will liaise with four Regional Operational Technical Units (selected based on ecological zones) that will be charged with the responsibility of collecting physical (meteorological, hydrological, atmospheric, pedological, and biological) and socio-economic parameters on the subject. The latter will be responsible for the follow-up of impacts and the elaboration of emergency strategies and responses towards the hazards posed by climate change to the environment and the population. However, the legal framework defining the functions of such an observatory will have to be defined.

Other approaches towards a renewed ethic on the phenomenon include awareness creation and sensitization of the community through capacity-building lectures designed to empower Members of Parliament and Councillors who constitute the policy-making and legislative arm of the country. Communications Guides on Climate Change have been prepared to facilitate the implementation of the sensitization process by these stakeholders.

The implementation of strategies designed by the United Nations Environment Programme (UNEP) to mainstream environment and sustainable development issues into tertiary institutions across the African continent has been effected in over seventy-two countries, including in Cameroon at the University of Buea. This involves incorporating climate change and sustainable development issues into a compulsory university course as well as into university teaching and research programs towards this goal (unep.org).

The fourth strategy on the continent is that adopted by the International Council for Science Regional Office for Africa (ICSU-ROA) in its Science Plan from 2007 to 2012 in which it has incorporated climate change-related projects such as Global Environmental Change including Climate Change and Adaptation, Natural and Human Induced Hazards and Disasters among others across the continent. The successful implementation of these Science Plans will constitute a renewed scientific ethic towards mitigation and redressing (or coping with) the effects and impacts of this phenomenon (icsu-africa.org).

Conclusion and Future Perspectives

An evaluation of the state, trends, and impacts of parameters of climate change in Cameroon is indicative of an ever increasing trend in global warming, accompanied by a decreasing but sporadic trend of high precipitation. The similarity of magnitudes of these trends with global patterns is indicative of the severity of this phenomenon as testified by the wide variety of its global and national impacts. Related disasters in the country show a constant increase with time, which clearly corroborates the global patterns of disasters caused by this phenomenon. The impacts in the country are, however, seen to be most severe within the Savannah and Sahel ecological zones, where droughts have been shown to have disastrous effects on agriculture and water resources, thereby destabilizing the socio-cultural and economic lifestyles of the communities leading to increasing numbers of environmental refugees.

This calls for urgent actions towards conceiving and implementing mitigation and adaptation approaches that could increase the resilience of these communities despite contradictory ideas by skeptics with respect to the threat from this phenomenon to the survival of humankind and other species on this planet.

The need for renewed environmental, scientific, and political ethics towards devising mitigation, adaptation, and resilience options and strategies is paramount. Within the Cameroonian and African contexts, steps have been taken to create an observatory for climate change with diverse functions that will include the definition and/or conception of approaches of harmonizing multidisciplinary efforts in collaboration with international collaborators on the issue, as well as sensitization and research approaches on the subject.

These approaches will only be effective if their implementation is holistic and cuts across the entire continent, since the impacts of the phenomenon do not respect territorial boundaries. This will therefore require a completely renewed political ethic similar to that of the ongoing "United Europe" process.

However, an overriding consideration for the effective success of these approaches will be based on the enactment of appropriate legislation as a fundamental tool for the implementation of policies and decisions. Projects concerning the theme should therefore come from a policy established by governments, based on a determined national strategy. Such a strategy should neither be "top-down" nor "bottom-up," but should be a combination of both, with the mutual participation of all. In this way, it will meet the real needs of the members of the community, who will equally contribute to the formulation of climate change strategies since their effective implementation will depend on local behaviors, especially at regional levels such as in agriculture and water resources management, which have been shown to be the most vulnerable in Cameroon and on the continent in general to this phenomenon.

References

Allen, Myles (2008) "Climate Progress: An insider's view of climate science politics and solutions." Retrieved from <http://climateprogress.org/2008/12/16>.

Ayonghe, Samuel (1998) "Environmental effects on groundwater of igneous aquifers in Cameroon." *Hydrology in a changing environment Vol. II,* British Hydrological Society, John Wiley and Sons, pp. 155-162.

Ayonghe, Samuel, Georges Mafany, Edwin Ntasin and Patric Samalang (1999) "Seismically activated swarm of landslides, tension cracks, and a rockfall after heavy rainfall in Bafaka, Cameroon." *Natural Hazards, 19*, 13-27.

Ayonghe, Samuel (2001) "A quantitative evaluation of global warming and precipitation in Cameroon from 1930 to 1995 and projections to 2060." *Readings in Geography*. Lambi C (ed.), pp. 142-155.

Ayonghe, Samuel, Edwin Ntasin, Patric Samalang and Emmanuel Suh (2004) "The June 27, 2001 landslides on volcanic cones in Limbe, Mt Cameroon West Africa." *Journal of African Earth Sciences*, 39, pp. 435-439.

Ayonghe, Samuel and Sanni Amawa (2007) "Human vulnerability to depleting water resources in Cameroon: Sensitisation approaches." *Southern African Journal of Environmental Education, 24*, 2007, 19-24.

Ayonghe, Samuel and Edwin Ntasin (2008) "The geological control and triggering mechanisms of landslides of 20th July 2003 within the Bamboutos Caldera, Cameroon." *Journal of the Cameroon Academy of Sciences, 7* (3), pp. 191-204.

Lagos, Ricardo (2008) "Le monde après 2012. " *International Agency for Atomic Energy (IAEA) Bulletin*, 49-2 March 2008, 16-19.

Lean, Geoffrey and Don Hinrichsen (1992) *Atlas of the Environment*. Helicon, 192. Retrieved from <www.alibris.com/search/books/isbn/0091774330>.

McBean, Gordon (2004) "Climate change and extreme weather: A basis for action." *Natural Hazards*, 31, pp. 177-190.

Nobel Laureates (2001) "The next hundred years." *Science*, 294, 5551.

Odjugo, Peter (2005) "An analysis of rainfall patterns in Nigeria." *Global Journal of Environmental Sciences*, *4*(2), 2005, pp. 139-145.

Pachauri, Rajendra (2008) "Le temps d'agir." *International Atomic Energy Agency (IAEA)*, Bulletin 49-2, March 2008, 12-15. Retrieved from <www.iaea.org/podcasts>. Sacchetti, Dana (2008) "Oser différer: Des avis dissidents sur les changement climatiques."

International Atomic Energy Agency (IAEA), Bulletin 49-2, March 2008, 20-21. International Council for Science, Regional Office for Africa. Retrieved from <www.icsuafrica.org>.

United Nations Environment Programme. Retrieved from <www.unep.org/training>.

14

Addressing Climate Change: Challenges, Ethics, and Hope

Taha Balafrej

Climate change, as a global and scientifically-proven phenomenon, is closely linked to local issues. This link is even more evident in developing countries, which are vulnerable to the effects of climate change while lacking the necessary means and resources for their adaptation. This paper attempts to show, by exploring the case of Morocco, and in particular, by highlighting the need to implement projects aimed at both adaptation and mitigation, that this global challenge provides important local opportunities for sustainable development. In this regard, water and energy sectors offer real possibilities for clean investment and technology transfer. By endorsing stronger ethics on domestic and global levels, there is great hope that this challenge can be met.

Introduction

The fight against climate change affecting ecosystems and humans raises the need for a renewed ethic. The challenge we are facing is not only a political issue. The selfish tactics and inertia that have characterized so far the negotiating process within the United Nations, as well as the general lack of commitment in favor of the poor and of future generations reveal the missing dimension: the ethical dimension. Truth, responsibility, solidarity, sharing, and equity are all fundamental principles of ethics we need to revive for the sake of our planet at the global level, but also locally and individually as well. This chapter aims at addressing this issue from a Moroccan point of view, based on some local experiences and perspectives.

In the United Nations grouping, Morocco is a member of the G77 and China Group. But like many others, Morocco is not an oil producing country. It is not a small island state, nor a big developing country, and it is not one of the least developed ones, but it is a country located in the corner of Africa, watching with envy and admiration towards Europe and America. The country is now at a crossroads. To survive, it needs to develop. And for that it must move forward, open up, and be attractive. To achieve this, it over-exploits its natural resources (water, forestry, fisheries, etc.), following what was supposed to be the model of countries that have advanced their people towards freedom, welfare, and stability. But resources are becoming scarce, the needs very high and the competitiveness ruthless, while Moroccan citizens suffer the brunt of the negative effects of climate change, although the country is a very low emitter of CO_2 in the atmosphere: only two tons per year on average, compared with twelve tons for Western citizens or with the three tons emitted by each long-distance flight passenger.

Worrying Truth

In a world that pays close attention almost exclusively to the news of the moment, which is attracted by spectacular scenes, the United Nations has succeeded in placing the issue of climate change at the top of the political and diplomatic agenda. In recent years, indeed, this global phenomenon has emerged as a matter of concern for world leaders, and international public opinion as well. But the impression is given that there is a yawning gap between, on one side, the discussions and resolutions, and on the other, the suffering of people, especially the poor, due to the disruption of the climate as it is experienced every day.

The pollution produced by factories in Illinois yesterday is responsible for adverse effects on the people of Chad today, and it will impact other parts of the world tomorrow. The Intergovernmental Panel on Climate Change (IPCC), awarded the Nobel Peace Prize in 2007, specifies in its Fourth Report, with an unprecedented degree of certainty, that human activities through emissions of greenhouse gases (GHG) are responsible for global warming (IPCC 2007).

Reports of scientists and experts show that climate change is now a proven phenomenon. Drained rivers and devastating floods, fish extinction, soil erosion, heat waves, desertification, and disease—these consequences are already here and are more especially experienced by poor people in developing countries.

In connection with these physical and natural impacts, consequences of another kind are emerging. In ten years, "climate refugees will be counted in millions, most of them victims of global warming," predict high-level leaders of the European Union. In the Moroccan cities, one regularly sees an ever-increasing number of sub-Saharan immigrants who have fled the harsh climatic conditions in their countries and are waiting to cross the Mediterranean to enter what they consider to be a European paradise.

On the economic front, concerns are many. The Stern Report, published in October 2006, showed that "if we do not act, the overall costs and risks of climate change will be equivalent to losing at least 5% of global GDP each year, now and forever. If a wider range of risks and impacts is taken into account, the estimates of damage could rise to 20% of GDP or more" (Stern 2007). This report goes one step further by putting in parallel the damage of climate change and wars: "Our actions now and over the coming decades could create risks of major disruption to economic and social activity, on a scale similar to the disruptions associated to the great wars and the economic depression of the first half of the 20th century."

Faced with this global phenomenon, the leaders of major industrialized countries proclaim their will to reduce in a "substantial" manner their GHG emissions. They even agree that new actions against climate change would be carried pursuant to the principle of "common but differentiated responsibilities and respective capabilities." However, these statements cannot overcome the skepticism prevailing so far among the public.

In reviewing the data on GHG emissions, many comments come to mind. When one has to worry about the continuing emissions increase in many countries of the world, one has also to be worried about low emission levels as an indicator of slow development in many other parts of the world. While emissions of GHG in developed countries as a whole continue to grow, and emissions from emerging countries explode under the effect of their race to catch up with the rich countries, those of the African continent grow very slowly. The Millennium Development Goals Report released by the UN in 2008 shows the trend in CO_2 emissions as indicated in the table (in billions of tons):

And even if emissions per unit of economic output fell by more than 20 percent in developed regions, and increased by 35 percent in Southeast Asia and 25 percent in North Africa, per capita emissions remain the highest in developed regions—approximately twelve tons of CO_2 per

	1990	2000	2005
Developed countries	10.8	11.6	12.0
Developing countries	6.7	9.6	13.1
Subs. Africa	0.5	0.5	0.7
Northern Africa	0.2	0.2	0.4

(United Nations 2008)

person per year, compared with about three tons in developing regions and 0.8 tons in sub-Saharan Africa.

The regions that pollute the least are not the ones which are least vulnerable. On the contrary. The Fourth IPCC Report is very clear in its conclusions describing the vulnerability of the African continent, summarized in four important points that are worth being reproduced here:

- By 2020, between seventy-five and 250 million people are projected to be exposed to increased water stress due to climate change.
- By 2020, in some countries, yields from rain-fed agriculture could be reduced by up to 50 percent. Agricultural production, including access to food, in many African countries is projected to be severely compromised. This would further adversely affect food security and exacerbate malnutrition.
- Towards the end of the twenty-first century, the projected sea level rise will affect low-lying coastal areas with large populations. The cost of adaptation could amount to at least five to ten percent of Gross Domestic Product (GDP).
- By 2080, an increase of five to eight percent of arid and semi-arid land in Africa is projected under a range of climate scenarios. (IPCC 2007).

This vulnerability is not a new finding. The text of the United Nations Convention on Climate Change (UNFCCC), adopted in 1994 and now signed by 192 countries, has already reported the particularity of the African continent in its Article 4.1 (e). (UNFCCC 2005).

In Morocco, climate data recorded in the region during the twentieth century indicate a warming during this century, estimated at more than 1° C, with the increase in dry years, from a dry year every ten years at the beginning of the century, to five to six years of drought in a decade now, as it is highlighted in the Initial National Communication of Morocco published in 2001 (Agoumi 2003, 4). These data are confirmed by the Fourth IPCC Report, which ranks the Mediterranean region among those most vulnerable to decreasing rainfall. The race for competitiveness and investment exacerbates this vulnerability. Large tourist operators and major food companies, acting as virtual exporters of water, do not hesitate to over-exploit natural resources in host countries.

Insufficient Answers

This IPCC Report draws attention to the rescue objective: the reduction of emissions by at least 50 percent by 2050, compared to 1990 levels.

At a time when negotiations are committed to the establishment of the post-2012 regime, developed countries find it increasingly difficult to meet their commitment to reducing their emissions under the Kyoto Protocol. The objective of the European Union to reduce its emissions by 20 percent before 2020 provokes hostility from the business community and from some leaders, sensitive to the 44 billion Euros that this commitment would cost each year between 2013 and 2020. This number is worth comparing to the hundreds of billions injected into private institutions to save the financial system.

To address the effects of climate change, the focus has most often been placed on mitigation by industrialized countries through the limitation of GHG emissions in the atmosphere. The adaptation, which aims to reduce vulnerability, as a local response to a global phenomenon, has drawn the necessary attention only in very recent years. But the financial resources required for adaptation are enormous and exceed what developing countries can mobilize through their own resources. Multilateral funds and other sources of funding for international cooperation are not sufficient to meet this need.

In August 2007, the secretariat of the UNFCCC published a report in which numbers were given on "the investment and financial flows needed for adaptation in 2030 in addition to the investment and financial flows needed under a situation in which there is no climate change" (UNFCCC 2007, 20). These requirements would amount to more than 30 billion dollars in 2030, as summarized in the following table in billions of dollars:

For example, official development assistance, amounting to 5.2 billion dollars per year in infrastructure of water, should increase by about 150 percent by 2030 only to cover extra expenses due to climate change in developing countries.

The funding problem is crucial, but more serious is the lack of capacities in developing countries to design field projects and for monitoring

Sector	Global	Developing countries
Agriculture, forestry and fisheries	14	7
Water supply	11	9
Human health	5	5
Coastal zones	11	5
Infrastructure	8–130	2–41

and assessment. Based on Article 12 .4 of the UNFCCC[1], and on National Communications from 125 developing countries, the Secretariat of the Convention presented at the Conference of Parties in Montreal in 2005, a summary of 469 projects spread over the seven sectors identified by the IPCC for the mitigation of GHG[2]. The Secretariat and other agencies would be well advised to consider what happened to these projects three years after their communication to representatives of all countries of the world. In the case of Morocco, only four of twenty-three projects identified have been implemented, by domestic means.

The projects listed in the action plans of developing countries do not always come from a portfolio established by the government based on a determined national strategy. They are usually the result of an exploration work carried out by experts and officials, confined to a supporting role, without much influence on the planning of government action. Under these conditions, funding by international agencies does not always meet the real needs of recipient countries. This situation is also explained by the fact that the negotiators from developing countries involved in the UNFCCC process are generally unrelated to those working on development cooperation and to those who manage projects on the ground. Finance ministers are rarely involved in the deliberations and, often, diplomatic or purely technical aspects take precedence over aspects of cooperation and implementation. The negotiation process in the UN framework is lengthy, complex, and costly, and requires technical skills and budgetary resources that are beyond the capacity of many developing countries. Thus, a change in some of the current methods and procedures is essential.

The inclusion in the agenda of the political leaders of sustainable adaptation projects depends on the visibility that these projects provide. This visibility is, for example, greater for the inauguration of a large piece of infrastructure, even with negative impact on the environment, than when launching a program for saving water, and thus reducing vulnerability. It is greater for the launching of a thermal power plant than for the announcement of an energy efficiency policy.

In developing countries, beyond the standards and controls, anarchic adaptation is in progress on the ground. To address the shortage of clean water, people are using wastewater for irrigation in order to maintain the supply of food, with all the risks that it entails to health. And while this anarchic reuse of wastewater in agriculture is practiced routinely, governments are facing many obstacles for the legal reuse of treated wastewater for landscape and golf courses. On the other hand, adaptation projects may be quite successful. In Morocco, in less than six years, Seventy projects have prevented the pollution equivalent to that caused by four million inhabitants by a German-financed fund dedicated to reducing industrial pollution (FODEP). Among these projects, six have provided four million cubic meters per year for irrigation or for reuse in process production, with an investment of only 10 million dollars with 60 percent supported by industry. But lacking continuity in management and awareness is a real threat for this kind of initiative, which is not part of a comprehensive and integrated adaptation policy. A similar situation prevails in the energy sector. The development of policies and measures favorable to the promotion of renewable energy in developing countries with strong potential for wind and sun, is delayed by interests lobbying for maintaining conventional energy primacy under the pretext of cost and technology needs.

There are also other barriers to overcome. Coordination, good governance, and ownership by the people are often lacking. Rural Moroccan people show some reluctance to the introduction of photovoltaic kits, requesting instead to be connected to the national electricity grid like their urban compatriots, and to be part of "modernity." The needed change for sustainable development should not be seen as an invitation to development on the cheap. By fostering local initiatives for adaptation in the sectors of water and energy, awareness could be raised about the nexus between sustainability and the legitimate aspiration for development.

Since the agreement by the Heads of State on the Millennium Development Goals in 2000, it is recognized that some progress has been made. But the ultimate goal of halving poverty by 2015 remains uncertain. The report published in June 2008 by the IPCC on Water and Climate Change shows how the pursued seven goals are tied directly or indirectly to climate change, particularly the issue of water, and demonstrates that their attainment is still dependent on more significant progress in these areas (IPCC 2008). This IPCC Report recommends "[i]mproved school attendance through improved health and reduced water-carrying burdens, especially for girls," illustrating a well known reality in Morocco, where young girls from rural areas do not go to school because of the absence of appropriate sanitation, and school dropout by girls in mountainous areas is due to their primary task of fetching wood.

While it is universally acknowledged that poverty and social issues are intimately linked to climate change, the climate dimension is not yet integrated into the public policy for development. The legitimate aspiration of the South for economic development and social progress is still leaving the environmental concern at a lower level in public priorities. Social policies promoted in developing countries are often diverted from their purpose and promote actions harmful to the environment. While funding butane gas with the aim of ensuring access to fuel for the poor for cooking and heating, the Moroccan Government has to address simultaneously the use of this incentive by big farmers for pumping water, at the expense of collective water resources. In the same vein, low quality diesel is heavily subsidized by the Moroccan Government in order to facilitate public transportation, but in doing so it generates urban pollution and health deterioration.

Hopeful Future

The Stern Report cited above makes a meaningful diagnosis of the situation: "climate change is the greatest market failure the world has ever seen, and it interacts with other market imperfections" (Stern 2007, viii). Since the release of that report, the failure is getting more complicated and the challenge could be summarized in one question: How are we going to reverse this failure, in times of market turmoil, economic crisis, poverty, and climate change?

In the fight against climate change, more and more voices are calling for a new global economy. The same Stern Report provides promising prospects for a new low carbon economy, for "action on climate change will also create significant business opportunities, as new markets are created in low-carbon energy technologies and other low-carbon goods and services. These markets could grow to be worth hundreds of billions of dollars each year, and employment in these sectors will expand accordingly" (Stern 2007, viii).

During his election campaign and after his inauguration, US President Obama highlighted the need for solving the economic crisis by, among other policies, promoting new ways of confronting the issues of energy and climate. This is a strong signal to all countries, especially developing countries, that they should engage in a similar re-thinking of their policies.

But now, action must follow statements. The leaders of the industrialized countries should not use the pretext of the current economic crisis to curb the necessary partnership between North and South. On the contrary, this partnership should be enhanced and transformed for green win-win opportunities. This path has been anticipated in the Kyoto Protocol by setting up the Clean Development Mechanism (CDM). This mechanism advocates for partnership between industrialized and developing countries with the participation of the private sector, providing at the same time resources for the Adaptation Fund.[3]

Through the CDM mechanism, it is expected that the projects registered so far would direct at least 18 billion dollars of investment towards developing countries. It is also expected that 100 billion dollars could be invested annually in developing countries if the international community reached an agreement urging industrialized countries to reduce their emissions by 60 to 80 percent by 2050.

The increasing number of projects that are registered at the CDM Executive Board show that the mechanism works as a levy for investment and as a tool for technology transfer. However, this mechanism is unfair in terms of geographical distribution. Those among developing countries that present the highest growth rate, with high potential of CO_2 emissions, are the most attractive for partners from the industrialized countries. Thus, Africa's share in this mechanism remains insignificant. Out of the nearly 1400 projects registered so far,[4] less than thirty are African. Among these projects, four are Moroccan, designed and implemented after a long period of awareness-raising and capacity-building. By selling carbon credits from two of these projects (a wind farm and a program of photovoltaic kits for rural electrification), the agency in charge of electricity in Morocco is provided with annual revenues of about one million dollars. Similarly, a landfill near the capital Rabat, rehabilitated with recovery of methane, could earn more than 300,000 dollars per year, which can be used for maintenance and operation of the landfill.

Prevent or cure? Mitigate or adapt? The fourth IPCC Report recommends combining both policies: "there is high confidence that neither adaptation nor mitigation alone can avoid all climate change impacts, however, they can complement each other and together can significantly reduce the risks of climate change" (IPCC 2007, 20).

Projects combining mitigation and adaptation could foster greater commitment from local people and would be more attractive to foreign investors. Areas where this combination could have beneficial effects are many. In the case of Morocco, some projects are beginning to emerge, including: desalination of sea water using renewable energy, forestation both for carbon sinks and to combat desertification, and biofuels when they are not competing with crops for food and are not harmful to soil and water resources. These kinds of projects are among the most eligible for CDM and should benefit from preferential treatment for their registration or for their purchase by developed countries of credits to which they are entitled.

But the achievement of such projects depends on the technical capabilities of host countries. Capacity building and technology transfer are the key issues for any equitable partnership between the North and the South.

The post-Kyoto regime should be judged on the progress achieved in this area. In the 2007 Bali Conference, an agreement was reached for the developing countries to report on the actions they take to reduce their emissions, while the developed countries support them in the fields of technology, financing, and capacity building in a measurable, verifiable, and reportable manner.

However, another challenge is that the least developed countries, those without oil resources, and small island states exposed to the threat of extinction, require greater and broader involvement in negotiations leading to the expected agreement in Copenhagen in 2009.

Conclusion

In the current context of crisis and desire for change, there is room for hope that the problems of underdevelopment, from poverty to illiteracy, could find the beginning of their solution in the global mobilization for a new society. I hope that this society will be based on solidarity and common care for our planet, and on civic involvement for daily behavior free of unbridled consumerism and cultural uniformity, where rich countries change their current patterns of consumption and production.

I hope that Morocco will try its utmost to become a proponent of this new society. It has to enhance its capacities on the human and governance levels. It also has to be part of a world believing in green economy. It has to get ready to acquire the appropriate technologies for energy efficiency and for reuse of water. It has to reduce its dependence on fossil fuels by benefiting from the opportunities of the carbon market. It will have to gradually replace its exports of agricultural products to the European Union by solar and wind energy that could be inexhaustibly produced in its sunny and windy areas. Morocco has to be a country that does not limit itself to attracting offshore projects but becomes a leader in innovative clean technologies. Morocco has to recognize that by contributing to the global effort in addressing climate change, it could also guarantee the right for its people to improve their lives, and the general well-being of society.

To achieve all this, Morocco has to be self-reliant. It has many assets for that. But it must also be able to attract the support of developed countries. Cooperation in the common interest is the global dimension of the needed renewed ethics.

Notes

1. "It is open to developing country Parties to propose projects for funding in specific technologies, materials, equipment, techniques or practices that would carry them out, and if possible give an estimate of all costs of these projects, progress in reducing emissions and removals of GHG and an estimate of benefits that can be expected."
2. Agriculture, Energy, Forest, Industry, Building, Sanitation liquid and solid, Transport.
3. The Fund, which was established under the Kyoto Protocol and designed in the Marrakech Accords in 2001, will receive a two percent share from CDM projects carbon revenues.
4. End of January 2009.

References

Agoumi, Ali (2003) *Vulnerability of North African Countries to Climatic Changes.* Adaptation and Implementation Strategies for Climate Change, IISD, available at http://www.cckn.net/pdf/north_africa_fr.pdf. Initial communication of Morocco available at: http://unfccc.int/resource/docs/natc/mornc1f.pdf.

Balafrej, Taha (2007) *Adaptation and Mitigation for Sustainable Development,* The Fifth Informal Meeting on Further Actions against Climate Change, MOFA, Tokyo, January 2007.

Balafrej, Taha (2008) "La protection de l'environnement peut rapporter, même au Maroc," *L'Economiste*, mars 2008.

Balafrej, Taha (2007) "Le changement climatique: menaces et opportunités, " *L'Economiste*, août 2007.

Intergovernmental Panel on Climate Change (IPCC) (2007) *Climate Change 2007: Synthesis Report.*

Intergovernmental Panel on Climate Change (IPCC) (2008) *Technical Paper VI. Climate Change and Water.*

OCDE (2008) *Aspects économiques de l'adaptation au changement climatique.*

Stern, Nicholas (2007) "The Economics of Climate Change: The Stern Review." Cambridge University Press.

United Nations Framework Convention on Climate Change (UNFCCC) (2005) "List of projects submitted by Parties not included in Annex I to the Convention" (Montréal, November 2005).

United Nations Framework Convention on Climate Change (UNFCCC) (2007) "Investment and Financial Flows to Address Climate Change."

United Nations (2008) "The Millennium Development Goals Report."

Contributors

Robin Attfield is professor of philosophy at Cardiff University, UK, where he teaches ethics, environmental philosophy, and philosophy of religion. His latest books are: *Environmental Ethics: An Overview for the Twenty-First Century*, *Creation, Evolution and Meaning*, and *The Ethics of the Environment*.

Samuel Ayonghe Born in Cameroon, Dr. Ayonghe is an environmental geologist with a Ph.D. from Imperial College, University of London, UK. His research interests are in geohazards, groundwater resources, and climate change. He has coordinated national and international research projects, collaborates with UNEP and ICSU, has twenty-six published articles, and is presently coordinator of an Environmental Science Resource Centre and is vice dean of the Faculty of Science at the University of Buea, Cameroon.

Taha Balafrej is a professor and consultant in the sustainable development area. He is the author of several articles and numerous presentations on sustainable development and climate change, and has previously held positions as director of the Moroccan Ministry of the Environment, focal point of Morocco to the UNFCCC and the GEF, head of CDM DNA, and manager of several cooperation projects.

J. Baird Callicott is Regents Professor of Philosophy and chair of the Department of Philosophy and Religion Studies at the University of North Texas. He is the co-editor-in-chief of the *Encyclopedia of Environmental Ethics and Philosophy* and author or editor of a score of books and author of dozens of journal articles, encyclopedia articles, and book chapters in environmental philosophy and ethics. Callicott has served the International Society for Environmental Ethics as president and Yale University as Bioethicist-in-Residence. Callicott is perhaps best known as the leading contemporary exponent of Aldo Leopold's land ethic and is currently exploring an Aldo Leopold Earth ethic in response to global climate change.

Andrew Brennan has been professor and chair of philosophy at La Trobe University, Melbourne, since 2006. Before that he was chair of philosophy at the University of Western Australia, Perth, having previously been reader in philosophy at the University of Stirling, Scotland. His most recent book is a co-authored introduction to logic, published by Continuum in 2005 and subsequently translated into Portuguese and Spanish. He and Y.S. Lo are the authors of *Understanding Environmental Philosophy*.

Sheila D. Collins is director of the graduate program in public policy and international affairs at William Paterson University. A political scientist, she is the author or co-author of five books and numerous book chapters, encyclopedia entries, and articles on public policy, globalization, social movements, and religion. She is book series editor for the Caucus for a New Political Science of the American Political Science Association, a member of the Editorial Board of *New Political Science*, serves as co-chair of the Columbia University Seminar on Full Employment, Social Welfare and Equity and is an associate of the Columbia Seminar on Globalization and Popular Struggles. She is a co-founder and serves on the boards of the National Jobs for All Coalition, the United Nations Association of New Jersey, and the Council on International and Public Affairs. In 1984 she was National Rainbow Director for the Jesse Jackson presidential campaign.

John Gutrich is an ecological economist and chair and associate professor of the Department of Environmental Studies at Southern Oregon University. His research interests include ecological and environmental economics, sustainable development, global climate change, watershed management, invasive species control, wetland ecology and management, and ecosystem modeling. His research efforts have addressed optimal forest management and the value of carbon sequestration in New Hampshire forests, the sustainable use of upland forested watersheds in Hawaii, invasive species control, valuation of non-market ecosystem goods and services, restoration wetland ecology, ecological and economic risks of marine transgenic organisms, and cost-effective economic approaches considering federal anti-degradation requirements for rivers.

Thomas Heyd teaches in the department of philosophy and in environmental studies at the University of Victoria, British Columbia, Canada. He has held positions as the co-chair of "Culture, Values and World Perspectives as Factors in Responding to Climate Change," Mobilising the Populace, *Climate Change: Global Risks, Challenges and Decisions* Congress at Copenhagen in March 2009 (http://climatecongress.ku.dk/). His publications include *Encountering Nature: Toward an Environmental Culture*, *Recognizing the Autonomy of Nature*, *Analyse der Bedingungen für die Transformation von Umweltbewusstsein in Umweltschonendes Verhalten* (co-author with Schluchter et al.), and *Aesthetics and Rock Art* (co-editor with John Clegg).

Y. S. (Norva) Lo received her Ph.D. in philosophy (with distinction) from the University of Western Australia in 2002 and has been a member of the philosophy department of La Trobe University since 2006. She has been published in journals such as *Inquiry*, *Proceedings of Aristotelian Society*, *Environmental Ethics*, and *Environmental Values*. Her research interests include moral philosophy, environmental philosophy and psychology, experimental philosophy, and David Hume. Among her more recent publications are "Making and Finding Values in Nature" (Inquiry 2006) and "Environmental Ethics" (co-authored, in Stanford Encyclopedia of Philosophy 2008).

Freya Mathews is an Australian philosopher and author. Her work is mainly concerned with ecological philosophy, but also deals with questions of metaphysics, epistemology, ethics, and politics, as well as a variety of themes such as cosmology, place, identity, and indigeneity versus modernity. Her books include: *The Ecological Self*, *For Love of Matter: A Contemporary Panpsycism*, *Journey to the Source of the Merri*, and *Reinhabiting Reality: Towards a Recovery of Culture*.

Dune Lankard is an Athabaskan Eyak from the Copper River Delta of Alaska. He was a commercial fisherman until March 24, 1989, when the Exxon Valdez tanker spilled more than thirty million gallons of crude oil into Prince William Sound. Dune became a community activist and social entrepreneur that day, dedicating his life to habitat preservation and environmental justice. With healthy wild salmon ecosystems at the heart of his work, Dune promotes preservation-based sustainable economies, and speaks to the long-term impacts of the Exxon disaster and environmental destruction. Dune has created five non-profit organizations, and is starting several sustainable for-profit entities, exemplifying the equation: non-profit + for-profit = social profit.

Ved P. Nanda (chief editor) is John Evans Professor at the University of Denver and Thompson G. Marsh Professor of Law at the University of Denver Sturm College of Law, where he also serves as director of the International Legal Studies Program. In 2006, Professor Nanda was honored with the founding of the Nanda Center for International Law. He was vice provost at the University of Denver from 1994 to 2008. He holds or has held numerous official posts in professional organizations, such as the World Jurist Association, International Law Association, American Society of International Law, the American Bar Association International Law Section, and the United States Institute of Human Rights. He was formerly the US Delegate to the World Federation of the United Nations Associations, Geneva. He is chair of the Board of Trustees of the Uberoi Foundation for Religious Studies and a member of the Board of Trustees of the Iliff School of Theology, Denver.

Among numerous national and international awards, in September 2005, Professor Nanda was awarded the World Jurist Association's Highest Order of Justice and its World Legal Scholar award. In 2004 he received the Gandhi-King-Ikeda Award for Peacebuilding from Soka Gakkai International and Morehouse College. He has received honorary doctorates from Soka University in Tokyo, Japan and from Bundelkhand University, Jhansi, India. He is widely published in law journals and national magazines, has authored or co-authored twenty-three books in the various fields of international law and over 180 chapters and major law review articles. He is a frequent guest on television and radio and writes regularly on international issues for the *Denver Post*.

Katrina S. Rogers is a historian and political scientist, and professor and associate dean for the doctoral program at Fielding Graduate University, Santa Barbara, California. She is also the director for Fielding's Institute for Social Innovation (ISI), which strengthens the capacity of individuals and organizations to address societal problems. Rogers has served the conservation sector, ranging from leading local initiatives to advocating for stronger environmental protection at the state and national levels. In addition to many articles, her book, *Towards a Postpositivist World*, articulated a hermeneutic model for understanding global environmental policy. Her service on boards includes the Environmental Fund of Arizona, the Santa Barbara Parks Foundation, and the Grand Canyon Community Council. She received a post-doctoral fellowship from the Humboldt Foundation, and was a Fulbright scholar to Germany. She is currently working on a book, "Soft Power and Global Environmental Problems."

Alicia Villamizar is an associate professor in the environmental studies department and is chief of the Natural Resources Institute at Simón

Bolívar University. She has spent twenty years working in environmental impact assessment (EIA). Her research is concentrated in the ecology and management of mangroves, their vulnerability to sea level rise, and environmental laws, and she has taught ecology of mangroves, human ecology, sustainable development, and environmental impact assessment. She is a visiting professor at the Institute of Groundwater, University of Oklahoma (1997); Rosenstiel School of Marine and Atmospheric Science, Miami University (2004); School of the South Frontier, Chiapas-Mexico (2005) and Center of Atmospheric Sciences, Universidad Nacional Autónoma de México (2006). She is the author and co-author of twenty scientific and technical papers in ecology and environment.

Laura Westra is professor emerita (Philosophy), University of Windsor, Ph.D. in Law, Osgoode Hall Law School, and she teaches environmental law at the University of Windsor, Canada. After her first Ph.D. in philosophy, she taught at several universities in Canada and the US, specializing in environmental ethics and environmental justice. In 1992, she founded the Global Ecological Integrity Group, and received funding from Canadian sources until 1999 and from NATO in 2000 in order to continue its work. In February of 2005, she received her second Ph.D. in jurisprudence. She is the author of nineteen books and monographs, and over eighty articles and chapters in books.

For Product Safety Concerns and Information please contact our EU representative GPSR@taylorandfrancis.com
Taylor & Francis Verlag GmbH, Kaufingerstraße 24, 80331 München, Germany

www.ingramcontent.com/pod-product-compliance
Ingram Content Group UK Ltd.
Pitfield, Milton Keynes, MK11 3LW, UK
UKHW022024061126
477531UK00002B/18

* 9 7 8 1 1 3 8 5 2 0 6 6 0 *